LEARNING RESOURCES CTR/NEW ENGLAND TECH.
GEN TK7895.M5 C367
Carr, Joseph   Designing microprocessor-b

3 0147 0000 8347 0

TK7895 .M5 C3
Carr, Joseph
Designing mic.
 instrumentation

S0-CJP-092

Designing Microprocessor-Based Instrumentation

Designing Microprocessor-Based Instrumentation

JOSEPH J. CARR

Reston Publishing Company, Inc., Reston, Virginia
A Prentice-Hall Company

Library of Congress Cataloging in Publication Data

Carr, Joseph J.
  Designing microprocessor-based instrumentation.

  Includes bibliographical references and index.
  1. Microprocessors. 2. Computer interfaces.
  3. Electronic instruments. I. Title.
  TK7895.M5C367   001.64   82-3812
  ISBN 0-8359-1270-1   AACR2

Copyright © 1982 by
RESTON PUBLISHING COMPANY, INC.
A Prentice-Hall Company
Reston, Virginia 22090

All rights reserved. No part of this book may be reproduced in any way, or by any means, without permission in writing from the publisher.

10 9 8 7 6 5 4 3 2 1

Printed in the United States of America.

Contents

PREFACE   ix

1 INSTRUMENTATION   1

1.1 Analog vs. Digital, 2   1.2 Including Analog Electronics, 5

2 INTRODUCTION TO MICROCOMPUTERS   6

2.1 Objectives, 6   2.2 Self-Evaluation Questions, 6   2.3 Introduction, 6
2.4 Definitions, 7   2.5 Advantages of Microcomputers/Microprocessors, 15

3 EXAMINATION OF TWO POPULAR uP CHIPS   19

3.1 Objectives, 19   3.2 Self-Evaluation Questions, 19   3.3 Introduction, 19
3.4 The Zilog Z80, 20   3.5 The 6502, 28

4 BINARY ARITHMETIC AND DIGITAL CODES   32

4.1 Objectives, 32   4.2 Self-Evaluation Questions, 32   4.3 The Decimal Number System, 32   4.4 Notes on Notation, 33   4.5 Binary (Base-2) Numbers, 34   4.6 The Octal (Base-8) Number System, 35   4.7 The Hexadecimal (Base-16) Number System, 36   4.8 Arithmetic Operations, 37   4.9 Two's Complement Notation, 38   4.10 Binary Coded Decimal (BCD), 39   4.11 Binary Coded Octal (BCO), 39   4.12 Binary Coded Hexadecimal (BCH), 41   4.13 Machine Codes, 42   4.14 Character Codes, 43

vi CONTENTS

5 INTERFACING TECHNIQUES 44

5.1 Objectives, 44 5.2 Self-Evaluation Questions, 44 5.3 Introduction, 44
5.4 Interfacing Different IC Logic Families, 45 5.5 Interfacing with LEDs, Lamps, Relays, and Other Devices, 47 5.6 Tri-State Devices, 50 5.7 Bus Drivers, 51
5.8 Using Optoisolators, 53

6 MICROPROCESSOR SUPPORT CHIPS 54

6.1 Objectives, 54 6.2 Self-Evaluation Questions, 54 6.3 Support Chips, 55
6.4 Z80-PIO, 55 6.5 Z80-SIO, 57 6.6 Z80-DMA, 61 6.7 Z80-CTC, 64

7 SOLVING PROBLEMS WITH THE MICROPROCESSOR 66

7.1 Programming Exercises, 71 7.2 Integration and Differentiation, 72

8 ADDRESS DECODERS 78

8.1 Objectives, 78 8.2 Self-Evaluation Questions, 78 8.3 Addressing in Microcomputers—A Review, 78 8.4 Eight-Bit Decoders, 79

9 INTERFACING MEMORY 88

9.1 Control Signals for Memory Operations, 88 9.2 Dynamic Memory, 91
9.3 Adding WAIT States, 92 9.4 Memory-Mapped Devices, 94

10 INTERFACING I/O PORTS 98

10.1 Objectives, 98 10.2 Self-Evaluation Questions, 98 10.3 I/O Ports, 98
10.4 Generating Port/Device Select Signals, 99 10.5 Parallel Output Port Design, 103 10.6 Parallel Input Port Design, 105 10.7 Bidirectional Parallel I/O Port Design, 105 10.8 Serial I/O Ports, 108

11 SIGNALS AND NOISE 126

11.1 Different Types of Signals, 126 11.2 Noise, 130

12 OPERATIONAL AMPLIFIERS 132

12.1 Objectives, 132 12.2 Self-Evaluation Questions, 132 12.3 Operational

CONTENTS

Amplifiers: An Introduction, 132   12.4 Properties of the Ideal Op-Amp, 133   12.5 Differential Inputs, 134   12.6 Analysis Using Kirchoff and Ohm, 135   12.7 Noninverting Followers, 137   12.8 Operational Amplifier Power Supplies, 139   12.9 Practical Devices: Some Problems, 141   12.10 DC Differential Amplifiers, 144   12.11 Practical Circuit, 147   12.12 Differential Amplifier Applications, 150   12.13 Integrators, 152   12.14 Differentiators, 153   12.15 Logarithmic and Antilog Amplifiers, 155   12.16 Summary, 157   12.17 Recapitulation, 158   12.18 Questions, 158   12.19 Problems, 158   12.20 References, 160

13 TRANSDUCERS 161

13.1 Objectives, 161   13.2 Self-Evaluation Questions, 161   13.3 Transducers and Transduction, 161   13.4 The Wheatstone Bridge, 162   13.5 Strain Gages, 162   13.6 Bonded and Unbonded Strain Gages, 166   13.7 Strain Gage Circuitry, 168   13.8 Transducer Sensitivity, 171   13.9 Balancing and Calibrating the Bridge, 172   13.10 Temperature Transducers, 174   13.11 Thermistors, 174   13.12 Thermocouples, 177   13.13 Semiconductor Temperature Transducers, 177   13.14 Inductive Transducers, 180   13.15 Linear Variable Differential Transformers (LVDT), 181   13.16 Position-Displacement Transducers, 183   13.17 Velocity and Acceleration Transducers, 185   13.18 Tachometers, 186   13.19 Force and Pressure Transducers, 187   13.20 Fluid Pressure Transducers, 188   13.21 Light Transducers, 190   13.22 Capacitive Transducers, 191   13.23 Summary, 194   13.24 Recapitulation, 194   13.25 Questions, 195   13.26 Problems, 196   13.27 Design Projects, 197   13.28 References, 198

14 SAMPLE & HOLD CIRCUITS 199

14.1 Objectives, 199   14.2 Self-Evaluation Questions, 199   14.3 Introduction, 199   14.4 Analog Switches, 200   14.5 Simple S & H Circuit, 201   14.6 Sample & Hold Errors, 205   14.7 Summary, 208

15 ANALOG REFERENCE CIRCUITS 209

15.1 Objectives, 209   15.2 Self-Evaluation Questions, 209   15.3 Introduction, 209   15.4 Zener Diodes, 210   15.5 Precision Operational Amplifier DC Reference Supplies, 212   15.6 Integrated Circuit Reference Sources, 219   15.7 Current Reference Sources, 220

16 INTERFACING KEYBOARDS, SWITCHES, AND DISPLAYS 223

16.1 Objectives, 223   16.2 Self-Evaluation Questions, 223   16.3 Keyboards, 224   16.4 Interfacing Keyboards to I/O Ports, 226   16.5 Interfacing Keyboards to Data Bus, 227   16.6 Interfacing Pushbuttons to the Microprocessor, 233   16.7 Summary, 239   16.8 Questions, 239

17 BASICS OF DATA CONVERSION—D/A 240

17.1 Objectives, 240   17.2 Self-Evaluation Questions, 240   17.3 What Are Data Converters?, 240   17.4 Binary Resistance Ladder Circuits, 241   17.5 R-2R Resistance Ladder Circuits, 244   17.6 Summary, 247   17.7 Recapitulation, 247   17.8 Questions, 247   17.9 Problems, 247   17.10 References, 247

18 BASICS OF DATA CONVERSION—A/D 248

18.1 Objectives, 248   18.2 Self-Evaluation Questions, 248   18.3 Introduction to Analog-to-Digital Conversion, 248   18.4 Types of A/D Converters, 249   18.5 Servo ADC Circuits, 249   18.6 Successive Approximation ADC Circuits, 251   18.7 Parallel Converters, 254   18.8 Voltage-to-Frequency Converters, 255   18.9 Dual-Slope Integration, 256   18.10 What Does the Dual-Slope DVM Measure?, 260   18.11 Summary, 261   18.12 Recapitulation, 261   18.13 Questions, 261   18.14 Problems, 262   18.15 References, 262

19 DATA ACQUISITION SYSTEMS 263

19.1 Objectives, 263   19.2 Self-Evaluation Questions, 263   19.3 Introduction, 263   19.4 Components of the Data Acquisition System, 265   19.5 DAS Selection, 269

20 DATA CONVERTER INTERFACING 270

20.1 Objectives, 270   20.2 Self-Evaluation Questions, 270   20.3 Data Converters and Data Conversion, 273   20.4 DAC Interfacing, 274   20.5 Interfacing the A/D Converter, 281   20.6 Summary, 288   20.7 Questions, 288

21 SOFTWARE DATA CONVERSION 289

21.1 Objectives, 289   21.2 Self-Evaluation Questions, 289   21.3 Introduction, 289   21.4 Software Ramp Converters, 290   21.5 Software Successive Approximation A/D Conversion, 293   21.6 Software Data Conversion Exercises, 296

Appendix A   Z80 INSTRUCTIONS SORTED BY OP-CODE   299

Appendix B   Z80 INSTRUCTIONS SORTED BY MNEMONIC   305

Appendix C   Z80/8080 INSTRUCTION EQUIVALENCY   313

INDEX   321

Preface

Five years ago engineering school professors were telling students that the microprocessor showed "substantial promise" for the future. Three years ago, the same professors were telling EE students that "within five years the EE who cannot design with microprocessors will be unemployable." Both projections have proven very much short of the mark! The microprocessor has shown itself to have far more than just "substantial" promise, and, indeed, in many areas of electrical engineering the working engineer must know microprocessor basics in order just to hold a job! In the electronic instrumentation field, and in the related control systems and process control fields that requirement is even tighter. Non-electrical engineers are also hopping on the bandwagon, and it is not at all unusual to find mechanical engineers, chemical engineers, biomedical engineers, naval architects, and others using microprocessors in instrumentation engineering applications. Often, these non-EEs will design their own hardware and software around single-board computers designed by the EE. It seems that, like the operational amplifier before it, the microprocessor has made the contriving of contrivances easier and broadened the spectrum of capable designers.

This text is basically an interfacing book with a flavor reminiscent of electronic instrumentation. I have included chapters on transducers and operational amplifiers because these devices are often used in microprocessor-based instruments as well as in their analog predecessors.

Joseph J. Carr, MSEE
Arlington, Virginia

1
Instrumentation

It has been claimed that man does not really understand any phenomena of the physical universe unless it can be *measured, controlled,* or *both*. There would be little point for the physician and the life scientist to even consider a concept such as blood pressure unless there were a means for quantifying, i.e., measuring, that pressure. Their efforts would be reduced to subjective guesses concerning the pressure and its meaning.

And what about *temperature?* We can subjectively index ranges of temperature by using relative terms like "hot" and "cold," but developing and producing modern devices that operate on the laws of thermodynamics require us to be able to quantify temperature. The Fahrenheit, Celsius and Kelvin temperature scales were devised to help us in this matter.

Most devices are considered "instruments" if they measure or control something. These instruments will examine some physical parameter and then display its value either in the form of numerics or as displacements on another instrument such as an oscilloscope or chart recorder. Or they will use the value in the calculation of some other parameter or control function.

Control systems are easily within the definition of "instrument." A physiologist researching glaucoma problems, for example, might use a servo-controlled perfusion apparatus to control the pressure in the eye of a test animal. Such an instrument monitors the pressure and the total volume of fluid perfused, and then makes calculations that drive the perfusion pump in a manner that keeps the pressure constant despite changes in the fluid volume of the eye. The list of possible instruments is seemingly endless.

It is convenient to classify electronic instruments according to type, i.e., simple display vs. calculating, and analog vs. digital.

A *simple display* instrument is one that takes an electrical signal that represents some physical parameter (i.e., a voltage analog—as in *analogous*—of that parameter) and displays it with no more processing, except possibly amplification or filtering. Examples of simple analog instruments include *voltmeters, ammeters,* certain *electronic thermometers,* certain *electronic manometers* (pressure meters), and so forth. If the instrument takes a raw signal, either "natural" or from a transducer, and applies it directly to the readout device with only amplification and/or filtering, then it is a simple readout device.

A *calculating instrument* must provide some signal processing that may include linear amplification, nonlinear amplification, scaling, filtering, logarithmic or anti-log amplification, integration, differentiation, level shifting, summation, multiplication, division, and so forth. A simple example of a calculating instrument is an electronic thermometer that produces an output signal that bears no numerical relationship to the desired units (°C, °K, or °F), and then produces a scale factor that has such a relationship, i.e., 10 mV/°K, with a scale that is selectable between °C and °F. The instrument must provide some amplification for scaling and possibly some level shifting.

Another example of a calculating instrument is the *cardiac output computer*. In this instrument, a thermistor probe is inserted into a blood vessel opening in a peripheral limb, and then threaded through the vessel to the pulmonary artery at the "output" of the right side of the patient's heart. The doctor then injects 5 cc or 10 cc of iced saline solution into the blood stream just prior to the heart, and the thermistor measures the temperature change at the other end of the heart. The computer then calculates the cardiac output by algebraically massaging the time integral of the temperature with factors such as the injectate volume, injectate temperature, blood temperature (preinjection), and some constants that are unique to the double-lumen catheter that carried the injectate into the heart. The result is computed, and then displayed for all to see.

1.1 ANALOG VS. DIGITAL

The early models of the cardiac output computer were all analog, but modern versions are all digital. This situation merely reflects the trend throughout the electronics instrument industry. What do these terms "analog" and "digital" mean, and are they always applied correctly? Also, are there any advantages to either analog or digital circuitry that make one clearly superior to the other as the choice of the practicing engineer?

Originally, the term "analog" referred to a voltage or current signal that represented some physical parameter. The voltage produced at the output of a bimetallic thermocouple, for example, is the *voltage analog of the junction temperature.* The word "analog" was used to denote some voltage that was *analogous* to some nonelectrical parameter. The word analog did not refer to the changing collector voltage on an audio amplifier transistor, but *it does now!*

By common usage, the term *analog* denotes any voltage or current that is continuous in both range and domain, whether or not the signal represents some physical parameter. While this may infuriate the purist who views language as static, it communicates effectively among engineers. The analog signal, then, is one that can take on any voltage or current value within its given range or domain at any time. There are no constraints placed on possible values except at the boundaries. A voltage, for example, may vary in some instruments from -10 to $+10$ volts, over a period of 0 to t—with no limits placed on either value.

A *digital signal,* on the other hand, is continuous in neither its range nor its domain. The range can take on only values that are representable by an integer number system (such as binary) and may exist only at certain discrete times in the domain. Where analog signals are often best represented as a plot on a graph, the digital signal can be "plotted" in the form of a table of values. Therein lies the ability to store a digital signal inside of a computer!

An analog instrument processes the signal in so-called "analog electronics" circuits, examples of which are amplifiers, differentiators, and others in the list given earlier. A digital instrument processes the signal in circuits such as gates, inverters, registers, and flip-flops. A modern dimension to digital instrumentation, and the subject of this book, is the use of *microprocessors* to replace many of the digital logic elements and some of the analog.

So what constitutes a "digital" instrument? It is the nature of the internal electronics, not just the display, that determines whether or not the instrument is truly "digital." Let's illustrate the point with two examples of electronic thermometers (Figure 1-1). All of these versions have existed.

In Figure 1-1(a) the temperature transducer produces an output voltage that is scaled at 10 mV/°K (a standard scale factor), and an amplifier is used to change the scale factor to 100 mV/°K. The output display device is a 0 to 19.99 volt digital voltmeter. To the outside observer, the instrument is a digital device, as evidenced by the digital display. The instrument might even be advertised as a "digital thermometer." But only the display uses digital electronics techniques. We would obtain an equally accurate readout by substituting an adequate analog voltmeter for the digital voltmeter. The instrument in Figure 1-1(a) is clearly an analog instrument, despite the fact that it uses a digital readout. While some advantage pertains to the use of digital displays, the instrument is not magically

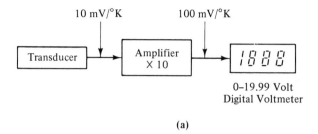

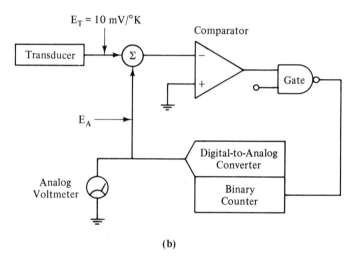

FIGURE 1-1. (a) Block diagram of a simple (analog) instrumentation system. (b) Block diagram of an analog-digital hybrid instrument system.

converted into a digital instrument by the replacement of one DC voltmeter for another.

Now consider the electronic thermometer shown in Figure 1-1(b). In this circuit, the transducer output is summed with an analog voltage from a digital-to-analog converter (DAC) to find the point at which the voltages are equal. The DAC output voltage is read on an analog voltmeter. Is this instrument analog or digital? It is neither; it is a hybrid of the two categories. If the display had been a digital counter, that single fact would have rendered the circuit "digital" instead of hybrid. In that case, the bulk of the internal circuitry is digital.

It is sometimes said that digital instruments contain certain built-in errors. This error is due to the digitization of the signal, and the display; i.e., error inherent in the process because only certain discrete values are allowed. When values are expressed as binary numbers, then the minimum resolution is set by the quantity represented by the least significant bit (1-LSB). This error is called the 1-LSB error. The implication is that analog circuits are more accurate and have better resolution because they will allow any value within the range. That concept is nonsense! Analog circuits contain substantial errors, including (but not limited to): *amplifier gain error, offset voltage* or *offset current error, drift error, reference potential error,* and others. The analog meter movement offers *inertia error* (it tends to integrate high frequency signals), *hysteresis error, calibration error,* and *parallax error.* There might also be error caused by the fact that the pointer does not cover just one value, but a tiny range of values.

In general, digital electronics neither guarantees nor prohibits accuracy over analog circuits. In many cases, however, the digital version of an instrument may be more stable than the analog version, especially if the measurement is sampled and then held for a period of time.

1.2 INCLUDING ANALOG ELECTRONICS

Engineers are fond of characterizing themselves according to the type of work they do, often according to the type of circuits they work with. In some laboratories and companies, you will hear engineers referred to as digital engineers, or analog engineers, or RF engineers, and so forth. But when you are designing an electronic instrument, you are an *instrumentation engineer,* and it behooves you to be familiar with all of the circuit techniques that might be useful in making the instrument work most effectively. In this book, we will introduce you to some of the techniques of designing with microprocessors and other digital devices; we will also include sections on the analog portion of the instrument. One must avoid a prejudice against analog electronics, because there might be a good reason for using an analog subsystem in an otherwise digital system. One reason might be that the digital approach to the solution of the problem might require far too much memory space or CPU time. A simple filtering algorithm might even work better than the analog equivalent, yet you will lack the resources to implement the digital version and stay under budget.

But lest some think that the author of a digital book is bending over backwards to favor the analog world, let me hasten to say that the principle advantage of the microprocessor is that it allows us to replace many analog *and digital* circuit functions with *software.* That one fact makes it very easy to modify the instrument or to provide for special contingencies required by specific users.

2
Introduction to Microcomputers

2.1 OBJECTIVES

1. Learn the difference between microprocessors and microcomputers.
2. Use microcomputer/microprocessors to perform elementary instrumentation chores.
3. State the different types of microcomputers available to the instrument designer.

2.2 SELF-EVALUATION QUESTIONS

Before studying the material in this chapter, try answering the questions given below. These questions test your knowledge of the subject matter. If you cannot answer a particular question, then place a check mark beside it and look for the answer as you read the text.

1. What is the difference between mainframe computers, minicomputers, and microcomputers?
2. What is the difference between a microcomputer and a microprocessor?
3. How can a larger minicomputer or microcomputer emulate a small process controller computer?
4. What *is* a process controller computer?

2.3 INTRODUCTION

The digital programmable computer has achieved an almost mystical reputation in the few decades since its invention. Professionals and laymen alike have stood in awe of the computer's capabilities. But recently computing has come down to earth and is now "a game for all." School children can now own a computer that is more powerful than the machine I used as an engineering student in the late '60s.

The costs of microcomputer equipment have dropped so much over the past few years that designers of electronic instruments, process controllers and other products can incorporate a 4.5 × 6.5-inch printed circuit board into their design—and that board will contain all the components of a microcomputer with up to 8K of memory and I/O ports! Computers are now used as components in other equipment designs.

2.4 DEFINITIONS

At one time definitions were a lot simpler. As a freshman engineering student at Old Dominion University (Norfolk, Va.) I was allowed to use an IBM model 1601 digital computer. Nobody had any doubt that the "1601" was a computer. But today a student (probably at the same institution) can sit at a small desk with a desk-top microcomputer complete with dual floppy discs, a video terminal, and a printer! In addition, the modern computer has more computing power than the old 1601 and costs less than one-tenth the mid-60's price of the old dinosaur (not counting the fact that 1967 dollars were bigger dollars).

Before attempting to discuss the design of electronic instrumentation that is microprocessor/microcomputer based, let's nail down some elementary definitions. Terminology tends to become a little sloppy in day-to-day use, both because of our own laziness and because galloping technology has tended to blur some of the distinctions that formerly served to delineate certain products (e.g., where is the dividing line between a small minicomputer and a large microcomputer?).

Two terms that always seem to be misused are *microprocessor* and *microcomputer*. Too often these terms are used interchangeably—even by me. In the paragraphs below we will discuss the differences between these and other terms.

Microprocessor. The microprocessor is a large scale integration (LSI) integrated circuit (Figure 2-1) that contains the central processing unit (CPU) portion of a programmable digital computer. The CPU section of a computer contains the arithmetic logic unit (ALU) that performs the basic computational and logical operations of the computer. The CPU also houses the control logic section that performs the housekeeping chores for the computer, and may have one or more registers for the temporary storage of data. All CPUs have at least one temporary storage register called the *accumulator,* or *A register.* The principle attribute of the

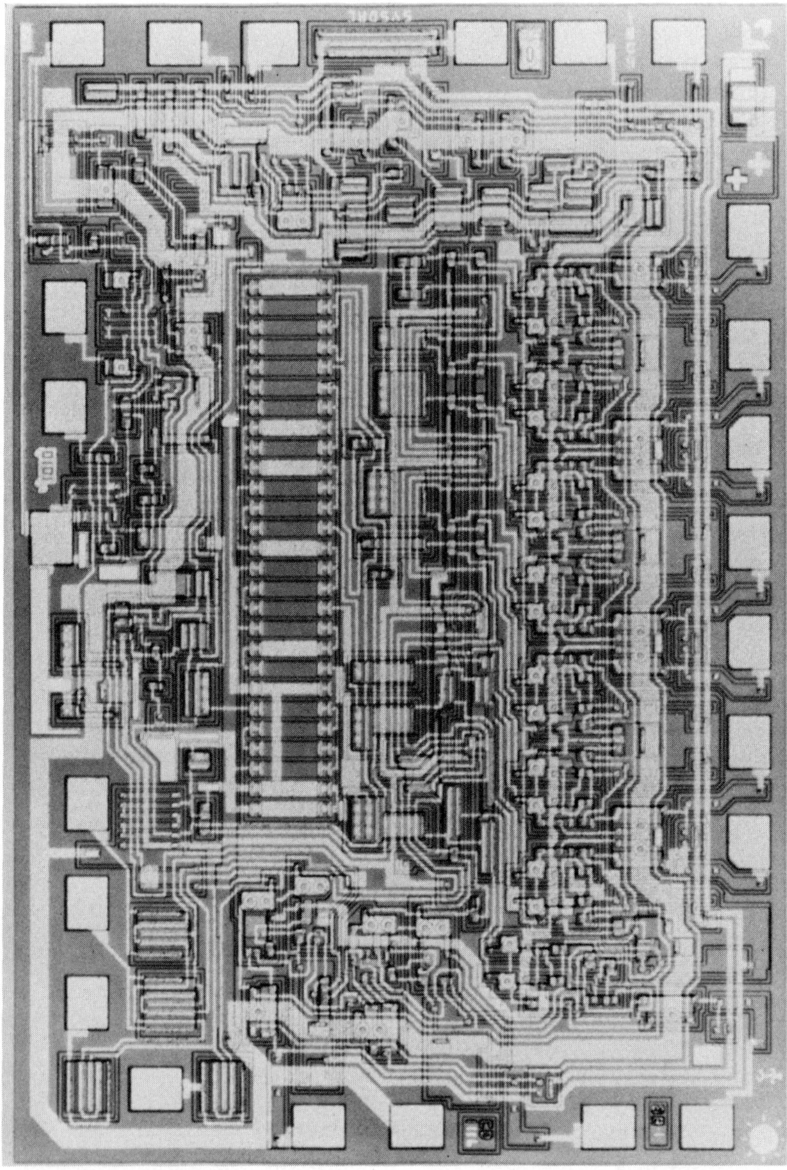

FIGURE 2-1. Microphotograph of an IC chip.

microprocessor is that it will execute instructions sequentially. The instructions are stored in coded binary form in an external memory. In general, a microprocessor will not contain any memory locations, except for the few internal registers mentioned above. To make the microprocessor into a microcomputer requires external memory—random access memory (RAM) and read only memory (ROM)—and external input/output (I/O) ports.

Microcomputer. A microcomputer is a full-fledged programmable digital computer built around a microprocessor integrated circuit. The uP acts as the CPU for the computer. In addition to the microprocessor chip, the microcomputer will have certain other external chips (as few as one or as many as 20) that provide external memory, I/O and other functions. The microcomputer may be as simple as the KIM-1, or as complex as a 21-board "S-100" system that is loaded with all the "goodies."

Single-chip computer. For several years we had no excuse for interchanging the terms microprocessor and microcomputer; a uP was an LSI IC, while the uC was a complete computing machine. But the 8048 and other devices began the process of dissolving the previously well-defined distinctions—it was both an LSI IC and a computer! A typical single-chip computer may have a CPU, two types of internal memory (RAM and ROM), and at least two I/O ports. Some versions of the 8048 family also contain built-in analog-digital converters so it may be directly interfaced with analog instrumentation and process control circuitry.

Single-board computers. The single-board computer (SBC) is a programmable digital computer, complete with I/O ports, certain "peripherals," and some memory, all on one single printed circuit board. Most of these devices are intended for use by students, either self-taught or classroom, or as small scale development systems for working professionals. Many users will make a single-board computer do for both the development and implementation of one-of-a-kind or small-scale projects. The SBC might have either a microprocessor or single-chip computer IC at its heart.

The "peripherals" on a single-board computer are usually of the most primitive kind (the Rockwell International AIM-65 and Ohio Scientific *Superboard II* are noteworthy exceptions), consisting of LED seven-segment displays and hexadecimal keypads much like the keypad on a handheld calculator or *Touchtone*® telephone. The typical display is capable of displaying only hexadecimal characters, and even these are a little optimistic, being constrained by the seven-segment LEDs (which were intended for decimal digits—the A-F digits of the hexadecimal system look a little funny). The Rockwell AIM-65, on the other hand, uses a regular ASCII keyboard and a 20-character 5 × 7 dot matrix display of LEDs. In addition, the AIM-65 has a 20-column dot matrix thermal printer on the board.

Most single-board computers have at least one interface connector that allows either expansion of the computer or interfacing into a system or instrument design. The manufacturers of most SBCs, such as the immensely popular KIM-1, probably did not envision their product in a wide application as a development system. The SBC is a *trainer* designed to introduce engineers and students to microcomputer technology at the minimum price.

Figures 2-2(a) through 2-2(c) show three popular single-board computers. The SBC shown in Figure 2-2(a) is the 6502-based SYM-1 by Synertek Systems, Inc. This device, which is not a kit, has the following features:

- 6502, eight-bit microprocessor
- 51 I/O lines
- Five interval timers
- keypad
- LED display
- 4K ROM resident monitor

FIGURE 2-2(a). *SYM-1* single-board computer. (*Courtesy of Synertek.*)

Section 2.4 / DEFINITIONS   11

FIGURE 2-2(b). *Superboard II* single-board computer. *(Courtesy of Ohio Scientific.)*

- 1K (optional up to 4K) RAM
- Up to 28K of user-provided ROM
- Audio cassette interface
- 20 mA TTY (full duplex)
- RS-232 interface
- Up to four relay drivers
- 32-character single-line oscilloscope display

The SYM-1 has a KIM-1 compatible bus, so it can be used in many applications that were once given to the KIM-1.

The SBC shown in Figure 2-2(b) is the Ohio Scientific *Superboard II*. Like the AIM-65, KIM-1, and SYM-1, the *Superboard II* is based on the eight-bit 6502 microprocessor chip. Like the AIM-65, this computer uses an ASCII keyboard. The keyboard makes the Superboard II capable of handling languages such as BASIC, which are impossible on a machine that has only a hexadecimal keypad. Still another SBC is shown in Figure 2-2(c). This machine is the Heathkit H-8, shown together with the H-17

FIGURE 2-2(c). *H-8* single-board computer-in-a-box. *(Courtesy of Heath/Zenith.)*

disc memory drive. The H-8 is a little more sophisticated as to packaging than some of the others shown in this section.

Minicomputers. Minicomputers predate microcomputers and were originally scaled-down versions of the larger machines. The Digital Equipment Corporation (DEC) model PDP-8 and PDP-11 are examples of "mini's." The minicomputer will use a variety of small-scale integration (SSI), medium-scale integration (MSI), and large-scale integration (LSI) integrated circuits to implement circuit functions.

Minicomputers have traditionally been more powerful than microcomputers. They had, for example, longer length binary data words (12 to 32 bits instead of 4 to 8 bits), and operated at faster clock speeds (6 to 12 mHz instead of 1 to 2 mHz common in microcomputers). But this is an area of fading distinctions! Digital Equipment Corporation, for example, offers the LSI-11 microcomputer that acts like a *mini*. Similarly, 16-bit microcomputers are now available as are devices that operate at over 6 mHz. It is now sometimes difficult to draw the line of demarcation between micro's and mini's when a Z80 based microcomputer is in the same size cabinet as a typical mini (and even boasts *hard* disc memory), while minicomputers can be bought in desktop cabinets without disc memory. Figure 2-3 shows a Heath H-11 microcomputer system that is based on the DEC LSI-11, and will run PDP-11 software. This tabletop mini/micro system comes complete with a dual-disc drive (H27) and a "smart" video display terminal.

Section 2.4 / DEFINITIONS

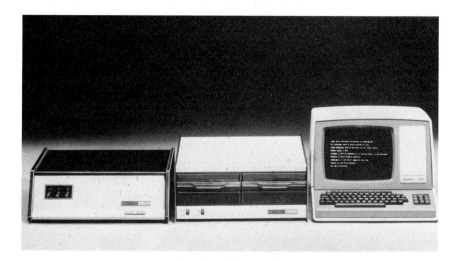

FIGURE 2-3. Small computer system with disc and video terminal. (*Courtesy of Heath/Zenith.*)

Larger microcomputers. To further blur the line between mini's and micro's, let us examine Figures 2-4 through 2-6. The computer shown in Figure 2-4 is the Heath/Zenith H89 machine, which is available in both kit form and assembled. It is an all-in-one computer that uses the Z80 microprocessor (the second Z80 described in the picture caption is part of the smart video terminal, which is the same as the H19). It contains a floppy disc memory drive, a full ASCII keyboard, spare numeric (hex) keypad, and 16K of RAM (expandable). All of the features are inside of the video display cabinet shown.

Figure 2-5 shows two versions of an S-100 *Altair* microcomputer. The S-100 designation has become generic, even though it was invented by the people at MITS who offered the *Altair* 8800 microcomputer. These 8800B models are front panel [Figure 2-5(a)] and turnkey [Figure 2-5(b)] models. The front panel sometimes assists in troubleshooting some programs and is useful for troubleshooting hardware problems in the computer. It will allow us to examine and modify any memory location without peripherals. It will also allow us to run the program either at full speed or on a single-step basis. The photo of the turnkey model shows the 20 S-100 printed circuit edge connectors into which the PC boards are plugged, Also shown is the power supply. This type of computer uses a +8 volt, unregulated, high current DC power supply. Each printed circuit board has its own +5 volt DC regulator, usually one to three one-ampere to three-ampere three-terminal IC regulators are used.

FIGURE 2-4. Heathkit H89 All-In-One Computer features two Z80 microprocessors, floppy disc storage, smart video terminal, heavy-duty keyboard, numeric keypad and 16K RAM—all in one compact unit. (*Courtesy of Heath/Zenith.*)

The popular *Apple II* microcomputer is shown in Figure 2-6. This tabletop computer will hold up to 48K of memory and has two 5.5-inch floppy disc drives and a video monitor. This computer takes up less tabletop space than a typewriter. The *Apple II* has a wide variety of plug-in accessory boards and will support sophisticated BASIC programs and a minicomputer-like operating system.

Mainframe computers. What comes to mind when most lay people think of "computers" is the mainframe computer. These machines are the computers used in large data processing departments. Microcomputerists, who have an elitist mentality based on the *smallness* of their machines, sometimes call mainframe computers "dinosaurs," and the programmers of such machines "dinosaur donkeys." But unlike their reptilean namesakes, these "dinosaurs" show no signs of extinction and are, in fact, an evolving species. The IBM 370 and CDC 6600 are examples of mainframe computers. Although there may be some competition between mini's, micro's, and mainframe computers in some applications, the typical micro does not in any way *directly* compete with the mainframe ma-

Section 2.5 / ADVANTAGES OF MICROCOMPUTERS/MICROPROCESSORS   15

FIGURE 2-5(a). MITS *Altair* 8800b—the original S-100 machine.

chine. The only competition is in the fact that microcomputers bring to small users the possibility of owning their own dedicated system; before, they were required either to use a time sharing system or send their computing chores out to a service vendor. The mainframe computer is almost irrelevant to the subject of this book, i.e., the design of microprocessor/microcomputer based electronic instrumentation.

2.5 ADVANTAGES OF MICROCOMPUTERS/MICROPROCESSORS

That microcomputers have certain advantages over other electronic circuit methods is attested to by the fact that so many are sold. The most obvious advantage is reduced *size*. Compared with dinosaurs, the typical micro is a mere lizard! An eight-bit microcomputer with 64K-bytes of memory can easily fit inside a tabletop cabinet. (See Figures 2-4 through 2-6.)

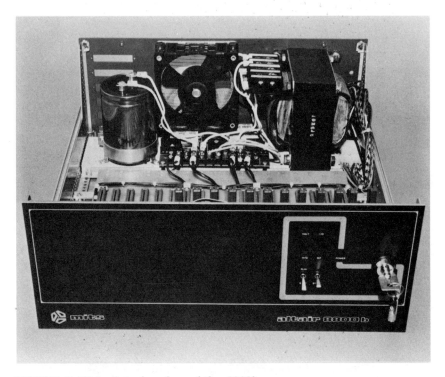

FIGURE 2-5(b). Interior view of the 8800b.

The LSI microcomputer chip is generally more complex than the discrete components circuit that does the same job. The interconnections between circuit elements, however, are much tinier (micrometers instead of centimeters). Input capacitances are therefore reduced. The metal oxide semiconductor (MOS) technology used in these devices produces very low current drains, hence the power consumption is reduced. A benefit of reduced power consumption is lower heat *generation* — so cooling and ventilation are reduced. A minicomputer might need a heavy-duty blower to keep the circuit cool, while a nearly equivalent microcomputer can use either a small blower or convection cooling.

The use of a few LSI chips (or just one) reduces the overall component count of the computer. Although this advantage relates to reduced size, it also functions to reduce the assembly cost of the final product and increases its reliability. The reliability aspect is due to the fact that the LSI computer chip seems as reliable as other forms of chip, of which dozens would be needed to replace the LSI chip. The reliability problem is especially improved in designs that use IC sockets. Some of the most invideous troubleshooting problems — both later in the field and more

Section 2.5 / ADVANTAGES OF MICROCOMPUTERS/MICROPROCESSORS

FIGURE 2-6. Small desktop computer with a lot of muscle. (*Courtesy of Apple.*)

immediately on the engineer's prototyping bench—are the result of defective IC sockets.

These advantages pertain to all microcomputer applications. But what are the advantages of the microcomputer or microprocessor to the designer of electronic instrumentation? One principal advantage is the fact that the drift of analog circuits is *eliminated*. Furthermore, changes in design or philosophy are not a cause for total redesign of the system in most cases. With the analog circuit, once the design is determined, it is final because of the difficulty in redesigning printed circuit boards. If an improvement is contemplated, or an extension of present capability is desired, or we find an unanticipated "glitch" in the design, then a complete redesign of the PC board may be necessary—unless we opt for the unprofessional and not very elegant solution called the *kluge board*. With the microprocessor-based design, however, many changes are possible with little or no change in the hardware: all changes are made in the software. Programs in uP-based instruments are typically stored in the form of read only memory (ROM) integrated circuits. These can be reprogrammed or replacement ROM chips with the corrected program can be substituted. This becomes an economical and easy job for the field service technicians—not the expensive and major overhaul required of many analog circuits.

3
Examination of Two Popular uP Chips

3.1 OBJECTIVES

1. To become familiar with the basic architecture of the Z80 device.
2. To become familiar with the basic architecture of the 6502 device.
3. To know the differences between the two devices.
4. To learn the different pinouts and control signals for the Z80/6502 devices.

3.2 SELF-EVALUATION QUESTIONS

Before studying the material in this chapter, try answering the questions given below. These questions test your knowledge of the subject matter. If you cannot answer a particular question, then place a check mark beside it and look for the answer as you read the text.

1. What is the purpose of the $\overline{\text{IORQ}}$ signal on the Z80 device?
2. What control signals are needed for an input operation on (a) the Z80 device, and (b) the 6502 device?
3. What is the length of the address word used with Z80 and 6502 devices?
4. How do the Z80 and 6502 devices treat input/output ports differently from each other?
5. Which of the two (6502 or Z80) uses memory-mapped I/O ports exclusively?

3.3 INTRODUCTION

There are numerous microprocessor chips on the market. The 6502 and Z80 devices were selected for the comparison made in this chapter because they represent two different basic philosophies in design and they are both immensely popular. There is a lot of available hardware and software for both Z80 and 6502 systems—something that cannot be said for some other selections.

3.4 THE ZILOG Z80

The Z80 is an integrated circuit microprocessor designed and manufactured by Zilog, Inc. (10460 Bubb Road, Cupertino, CA 95014), and second-sourced by Mostek, Inc. (1215 West Crosby Rd., Carrollton, TX 75006). The Z80 is similar to, but advanced over, the Intel 8080 microprocessor.

If you are familiar with the 8080 device, then making the switchover to Z80 will be very easy. The Z80 instruction set contains all of the 8080 instructions, plus a few more. It is usually claimed that the Z80 device has 158 different instructions, as opposed to only 78 for the 8080.

In general, any program that will run on an 8080 system, with the exception of those dependent upon timing loops, will also run on a Z80 system. There are differences in the clock timing, so those programs that create, or are dependent upon, specific 8080 timing will not usually run properly on the Z80.

Besides the different instruction set sizes, there are other differences between the Z80 and the 8080. The programmer of the Z80 device can use more internal registers and has more addressing modes than does the 8080 programmer.

In addition, there are several hardware differences. For one, the Z80 does away with the two-phase clock of the 8080. In the Z80, then, only a single-phase clock is used. The Z80 clock operates at 2.5 mHz, while the faster Z80A device will accept clock speeds to 4 mHz. The Z80 also differs from the 8080 in that it will operate from a single +5-volt power supply. The 8080 devices requires, in addition to the +5-volt supply, a −5-volt supply and a +12-volt supply.

The Z80 also provides an additional interrupt and the logic required to refresh dynamic memory.

The Z80 uses n-channel MOS technology, so it must be handled with care in order to avoid damage from static electricity discharge.

Figure 3-1 shows the block diagram to the internal circuitry of the Z80 device. Note that the Z80 contains the following sections: arithmetic logic unit (ALU), CPU registers, and instruction register, plus sections to decode the instructions received and control the address placed on the address bus.

The Z80 uses an eight-bit data bus and a sixteen-bit address bus. The

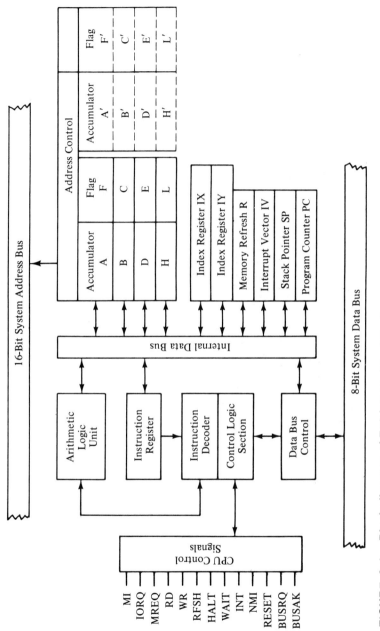

FIGURE 3-1. Block diagram of Z80 internal architecture.

use of sixteen bits on the address bus means that the Z80 can address up to 65,536 different memory locations.

The internal registers of the Z80 represent 208 bits of read/write memory that can be accessed by the programmer. These bits are arranged in the form of eighteen 8-bit registers, and four 16-bit registers. Figure 3-2 shows the organization of the Z80 register set.

The main register set consists of an accumulator (register A) and a flag register (register F), plus six general-purpose registers (B, C, D, E, H, and L). An alternate set of registers is provided that duplicates these registers: accumulator (A') and flag register (F'), plus the general-purpose registers B', C', D', E', H', and L'. Only one set of these registers can be active at any one time. One cannot, for example, use the B and B' registers without first using one of the instructions that interchanges the register sets.

The general-purpose registers can be paired to form three register pairs of 16 bits each: BC, DE, and HL. The alternate registers are also paired to allow 16-bit register pairs BC', DE', and HL'.

The Z80 special-purpose registers include interrupt vector I and memory fresh R (both 8-bit registers), and four 16-bit registers: index register IX, index register IY, stack pointer SP, and program counter PC.

Interrupt vector I. The I register is used to service interrupts originated by a peripheral device. The CPU will jump to a memory location

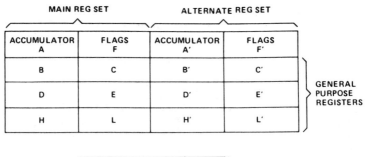

FIGURE 3-2. Register organization.

Section 3.4 / THE ZILOG Z80

containing the subroutine that services the interrupting device. The device will supply the lower-order eight bits of the 16-bit address, while the I register will contain the high-order eight bits of the address.

Memory refresh R. This register is used to refresh dynamic memory during the time when the CPU is decoding and executing the instruction fetched from memory. Seven bits of the R register are incremented after each instruction fetch, but the eighth bit remains as programmed through an LD R,A instruction. During refresh, a refresh signal becomes active, the contents of the R register are placed on the lower eight bits of the address bus, and the contents of the I register are placed on the upper eight bits of the address bus.

Index registers IX and IY. These registers are used to point to external memory locations in indirect addressing instructions. The actual memory location addressed will be the sum of the contents of an index register and a displacement integer d (or, alternatively, some instructions use the two's complement of d). Both IX and IY index registers are independent of each other. Note that many microprocessor chips do not have index registers at all.

Stack pointer (SP). The stack pointer is a two-byte register that is used to hold the 16-bit address of a last-in-first-out (LIFO) stack in external memory. The data to and from the memory stack are handled through the PUSH and POP instructions, respectively.

Program counter (PC). The program counter in any computer holds the address of the instruction being fetched from memory. In the Z80, the program counter is a 16-bit register. The PC will be automatically incremented the correct number of digits after each instruction (e.g., one-byte instructions increment PC + 1, two-byte instructions PC + 2, etc.). When a JUMP operation occurs, the program counter will contain the address of the location to which the program jumped. When it is RETURNED, the PC will contain the address of the next sequential instruction that would have been fetched if no jump had occurred.

Figure 3-3 shows how the program counter would work on a jump operation. Let us say that we have a program that starts at location 02 00 (hex), and finishes at location 02 06. But when it encounters the instruction at 02 02, it is an unconditional jump to location 06 12. Now, for the purposes of illustration, our subroutine at 06 12 is a RETurn instruction (useless in the real world, perhaps, but useful for illustration). It then jumps back to the next sequential location 02 05. Note that the next sequential location from 02 02 in this case is not 02 03, but 02 05. This is due to the fact that the jump instruction was a three-byte instruction. We had

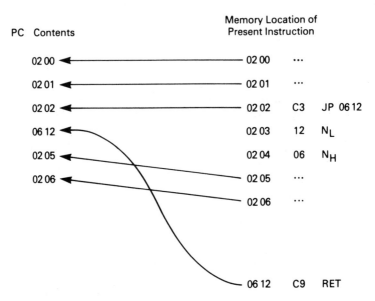

FIGURE 3-3. How the program counter works on a jump operation.

to give it the instruction (02 02), the low-order byte of the memory location to jump to (02 03), and the high-order byte of the memory location (02 04).

Arithmetic Logic Unit (ALU)

The heart of any computer or microprocessor, and the factor that distinguishes it from all other digital electronic circuits, is the arithmetic logic unit, or ALU. This circuit performs the data manipulation for the device. The functions possible in the Z80 uP are add, subtract, compare, logical AND, logical OR, logical exclusive-OR (XOR), left shift (logical), left shift (arithmetic), right shift (logical), right shift (arithmetic), increment, decrement, set a bit (i.e., make it 1), reset a bit (make it 0), and test a bit to see whether it is 1 or 0.

Flag Resisters (F and F')

The Z80 provides two status registers: F and F'. Only one is active at any one time, depending upon whether the programmer has selected the main register bank or the alternative register bank. These registers are each

Section 3.4 / THE ZILOG Z80

eight bits long and each bit is used to denote a different status condition. As a result, these bits of the F and F' register are also called *condition bits*.

The flags in the F and F' register are SET or RESET after certain arithmetic or other operations upon data. The program can then tell something about the result of the operation. The allocations are as follows:

BIT (F/F')	DESIGNATION	MEANING
0	C	Carry flag. Indicates a carry from the high-order bit of the accumulator (B7).
1	N	Subtraction flag used in BCD subtract operations.
2	P/V	Parity/overflow
3	X	Undetermined
4	H	BCD *half-carry* flag (bit 4 in BCD operations)
5	X	Undetermined
6	Z	Zero flag is SET if the result of an operation is zero.
7	S	Sign flag is SET if the sign of a result after an operation is negative, RESET if it is zero or positive.

Z80 Pinouts

The Z80 device is constructed in a standard 40-pin DIP integrated circuit package. Since the Z80 uses NMOS technology, one is cautioned to become familiar with the rules for handling such devices before trying to handle the Z80 device. Those rules are actually very simple, so failing to follow them will net you what you deserve—a zapped IC.

Figure 3-4 shows the Z80 pinouts and package configuration. The definitions of the pinouts are given below.

A0-A15   Address bus (16 bits). Permits addressing up to 64K (i.e., 65,536 bytes) of memory, plus 256 different I/O ports. The address bus is active when HIGH, and has tri-state outputs. The entire 16 bits are used to address memory, while only the low-order byte (A0-A7) is used to address I/O ports.

D0-D7   Eight-bit data bus terminals. The data bus is, like the address bus, active high and uses tri-state outputs.

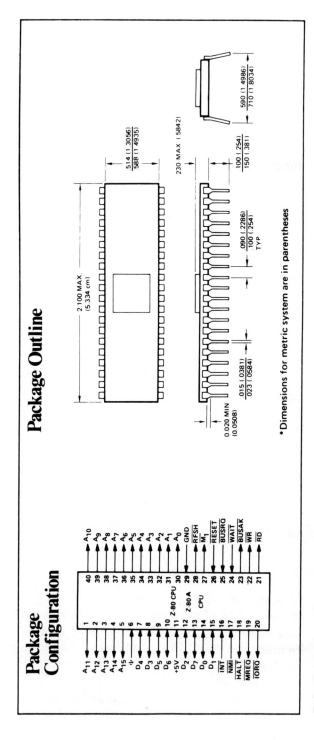

FIGURE 3-4. Z80 package and pinouts.

Section 3.4 / THE ZILOG Z80

$\overline{M1}$	Machine cycle 1. When this terminal is LOW, the CPU is in the op-code fetch portion of the instruction/execution cycle.
$\overline{MREQ}$	Memory request signal. Is active low, and is an active low output. When this terminal goes low, the address on the address bus is valid for a memory operation (read or write).
$\overline{IORQ}$	Input/output request. This active low, tri-state output indicates that an I/O operation is to take place. The low-order byte of the address bus (A0-A7) contains the address (0-255) of the selected port. The contents of the accumulator may be placed on the high-order byte of the address bus during this period. The $\overline{IORQ}$ is also generated to acknowledge an interrupt request, and tells the interrupting device to place the interrupt vector word on the data bus (i.e., low-order byte of the address of the interrupt service program.)
$\overline{RD}$	This is an active, low, tri-state output that indicates when a read operation from memory, or an I/O device, to the CPU is taking place.
$\overline{WR}$	Tri-state, active, low output that indicates when a write operation from the CPU to a memory location, or I/O device, is taking place. Tells the memory or I/O device that the data on the data bus are currently valid.
$\overline{RFSH}$	Refresh signal. This is an active low output that indicates that the lower seven bits of the address bus contain a refresh address for the dynamic memory.
$\overline{HALT}$	Active low output that indicates that a halt instruction is being executed. The CPU executes NOPs while in the halt state, and is awaiting the receipt of an interrupt signal.
$\overline{WAIT}$	Active low input that indicates that the addressed memory, or I/O device, is not yet ready to transfer data to the data bus.
$\overline{INT}$	Active low input that tells the CPU that an external device has requested an interrupt. The CPU will honor the request at the end of the current instruction cycle, if the interrupt flip-flop (software controlled) is SET.
$\overline{NMI}$	Active low input for nonmaskable interrupt operation. This line will cause the CPU to honor the interrupt at the end of the current instruction cycle, regardless of the

	state of the interrupt flip-flop. Forces automatic restart at location 00 66 (hex).
$\overline{\text{RESET}}$	Active low input that enables the interrupt flip-flip, clears the program counter (i.e., loads PC with 00 00) and clears I and R registers. This terminal can serve as a hardware jump-to-00-00 control.
$\overline{\text{BUSRQ}}$	Active low input that requests that the CPU address bus, data bus, and the control signals go to the high impedance (tri-state) state so that some other device can obtain control of these buses. The $\overline{\text{BUSRQ}}$ has a higher priority than $\overline{\text{NMI}}$, and is always honored at the end of the present instruction.
$\overline{\text{BUSAK}}$	Active low output that is used with the bus request signal, and tells the requesting device that the CPU buses are now in the high impedance state. When $\overline{\text{BUSAK}}$ drops low, then the requesting device may take control of the buses.
Φ	Clock signal input. Wants to see TTL level at 2 mHz or 4 mHz (Z80A) maximum.
GND	DC and signal ground terminal.
+5	Power supply terminal, to which is applied +5 volts dc from a regulated power supply.

3.5 THE 6502

The 6502 microprocessor chip is available from the originator, *MOS Technology, Inc.*, and more than 15 secondary sources. Among the secondary sources are Synertek and Rockwell International who make the SYM-1 and AIM-65 microcomputers, respectively. The 6502 device is widely used in microcomputer systems as well as in small-scale process controllers and other similar applications. Figure 3-5 shows the block diagram of the 6502 architecture, while Figure 3-6 shows the pinouts.

That there are certain similarities between the Z80 and 6502 devices testifies only to the fact that they are both microprocessor chips. The Z80 and 6502 devices are designed to different philosophies, which are reflected in their respective internal architectures. The Z80 allows separate input/output commands. The lower order byte of the address bus will carry the port number (256 different ports numbered 000 to 255) address, while the high order byte of the address bus carries the contents of the accumulator that is to be output. The I/O data is fed to and from the accumulator over the data bus. The 6502, on the other hand, uses a memory

Section 3.5 / THE 6502

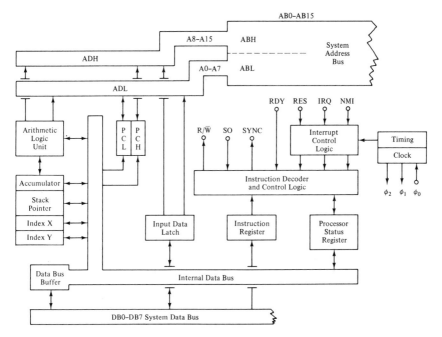

FIGURE 3-5. Block diagram of 6502 internal architecture.

mapping technique in which each I/O port is designated as a separate location in memory. We can then read or write to that memory location, depending upon whether the operation is an input or an output. The 6502 also lacks the multiple internal registers of the Z80. But this feature, like the lack of discrete I/O ports, does not hinder most microcomputer designs. Very few microcomputers will need more than a total of a dozen or so I/O ports and/or registers. Also, very few microcomputers will need the entire 64K (i.e., 65,536 bytes) of available memory addresses. In fact, most systems have less than 48K of memory. This allotment of memory would leave all locations above 48K for "firmware" (i.e., ROM) programs and I/O port or register selection.

Blocks in Figure 3-5 that have names similar to blocks in the Z80 diagram perform roughly similar (sometimes exactly identical) jobs for the 6502 device. Note, however, that the *program counter* (PC) is divided into two eight-bit (1 byte) registers called PCL (for low order byte) and PCH (for high order byte of 16-bit address). Similarly, the address bus is divided into low (ADL) and high (ADH) order segments. Unlike the Z80, the 6502 uses a multiphase clock for timing of the operations.

The 6502 pinouts are shown in Figure 3-6 and are defined below. Some of them are similar to Z80 pinouts, while others are unique to the 6502.

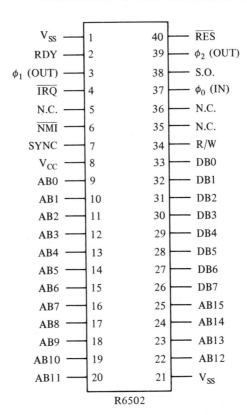

FIGURE 3-6. 6502 pinouts.

DB0–DB7	Eight-bit data bus
AB0–AB15	Sixteen-bit address bus
Φ_0	CPU clock
Φ_1, Φ_2	System clocks
R/$\overline{\text{W}}$	Indicates a *read* operation when HIGH, and a *write* operation when LOW. The normal inactive condition is read (i.e., HIGH). The CPU is writing data to the data bus (DB0–DB7) when this terminal is LOW.
$\overline{\text{IRQ}}$	*Interrupt Request.* This active-LOW input is used to interrupt the program being executed so that a subroutine can be executed instead. This interrupt line is maskable so it will respond only if the internal interrupt FF is enabled.

Section 3.5 / THE 6502   31

NMI   *Nonmaskable Interrupt.* Similar to the interrupt request, except that this active-LOW input cannot be disabled by the programmer.

RESET   Active-LOW *reset.* Essentially a hardware jump to location 0000_{16} instruction. If this terminal is brought LOW, then the PC is loaded with the address 0000_{16} and program execution starts from there. Can be used for manual or power-on reset operations, and does not alter the contents of the accumulator.

RDY   The *ready* signal is an input that will insert a wait state into the normal machine cycle sequence. The RDY line must make a negative-going (i.e., HIGH-to-LOW) transition during the Φ_1-HIGH clock cycle during any operation other than a *write*.

SO   *Set Overflow Flag.* This input will set the overflow flag *if* it makes a negative-going (i.e., HIGH-to-LOW transition during the trailing edge of the Φ_1 clock cycle).

SYNC   Active-HIGH output that is used to indicate the *instruction fetch* machine cycle.

Status Register (6502)

Like the F and F' registers in the Z80, the 6502 device has a status register that can be used by the programmer in a number of important ways. The status register can be read by the problem during certain operations, but is inaccessible to external hardware. The register has eight bits defined as follows:

BIT	USE
0	C—Carry. Indicates that a carry occurred from the accumulator on the last instruction executed. Active HIGH (logical-1)
1	Z—Zero. Indicates that the last operation performed resulted in a zero (if Z = HIGH) or nonzero (if Z = Low).
2	(Unused)
3	D—Decimal Mode. When this bit is HIGH it causes a decimal add with a carry or subtract with borrow for BCD operations.
4	B—Break. HIGH when an interrupt is executed.
5	(Blank)
6	O—Overflow
7	S—Sign. Indicates positive or negative results.

4
Binary Arithmetic and Digital Codes

4.1 OBJECTIVES

1. To learn the binary, i.e. base-2, number system.
2. To perform elementary calculations on binary numbers.
3. To use hexadecimal and octal numbers.
4. To learn ASCII, Baudot, and other character codes.

4.2 SELF-EVALUATION QUESTIONS

Before studying the material in this chapter, try answering the questions given below. These questions test your knowledge of the subject matter. If you cannot answer a particular question, then place a check mark beside it and look for the answer as your read the text.

1. What is the *decimal* equivalent of the binary number 11010011_2?
2. Find the sum of 1101_2 and 0100_2.
3. How would 11010011_2 be represented in hexadecimal? In BCD? In split-octal?
4. Why are ASCII character "7" and hexadecimal "7" given different binary representations?

4.3 THE DECIMAL NUMBER SYSTEM

The most familiar numbers system to man is the *decimal*, or *base-10*, numbers system. This is probably because of man's biquinary system of two sets of five fingers—five for each hand. When man started counting, it is probable that he used his fingers to make the count. Even today, the same word ("digits") is used to denote both the elements of the number system (i.e., 0, 1, 2, 3, . . . 9) and the fingers of the hand.

There are 10 digits of the decimal number system: 0, 1, 2, 3, 4, 5, 6, 7, 8 and 9. If we had to rely on only these ten digits at their so-called "natural" value, then technology would have remained in the primordial state in which man swapped and bartered using only low-valued integer numbers: e.g., four figs for a bowl of rice.

But technology did not remain primitive, and a major contributing factor was the introduction of the concept of a *weighted numbers system.* In a weighted system, a digit carries additional value by virtue of its *position* relative to other digits of the same number. The weights accorded to position in the decimal numbers system follow the rule:

$$a_n 10^n + \cdots + a_3 10^3 + a_2 10^2 + a_1 10^1 + a_0 10^0$$

or, written another way,

$$a_n 10^n + \cdots + (a_3 \times 1000) + (a_2 \times 100) + (a_1 \times 10) + (a_0 \times 1)$$

Where the *a* terms are the *digits* occupying that position. For example, consider the decimal number "432." Because our numbers system uses weighted value, we understand "432" to mean:

$$(4 \times 100) + (3 \times 10) + (2 \times 1)$$

or $\qquad (400) + (30) + (2) = 432$

In this case, $a_2 = 4$, $a_1 = 3$ and $a_0 = 2$

We have discussed the decimal numbers system in preparation for our discussion on numbers systems of other bases, especially base-2, base-8 and base-16. The base-2, or binary numbers system, is the "natural" system used in computer technology. The base-8 and base-16 systems are also used in computers, especially in the programming documentation and where the machine must deal directly with humans. The same concepts apply to all four numbers systems, so they are presented first in decimal in order to take advantage of the familiarity that you have with base-10 numbers.

4.4 NOTES ON NOTATION

Since we are going to use several different numbers systems in this book, we will need some notation conventions to differentiate the various numbers systems. For base-2, we will use the subscript "2." For example, base-2 "11" will be written in the form "11_2" to prevent confusing it with "11_8," "11_{10}," or "11_{16}." Similarly, we will use the subscript "8" for *octal* and "10" for *decimal* numbers.

Various sources use different notational methods to denote the base-16 *hexadecimal* numbers system. Some use the subscript "16" or "h," after *h*exadecimal. In other cases, a capital "H" will be used as either a prefix or suffix to the number. For example, "37_h," "37_{16}," "H37," and "37H" all denote the number hexadecimal "37." Some authors use special characters such as the "#" sign, or "$" sign to denote hexadecimal numbers. Thus, 37_{16} might also be found in some books or programmer's documentation in the forms "#37" or "$37." In this book, we will use subscript *16* and suffix "H" to denote base-16 hexadecimal numbers.

4.5 BINARY (BASE-2) NUMBERS

The binary numbers system is exactly like the base-10 numbers system—if you are missing eight fingers! Only two digits are used in the base-2, binary, numbers system: 0 and 1. Like the decimal system, the binary system is weighted by the position of the digits relative to each other. The rule is:

$$a_n 2^n + \cdots + a_3 2^3 + a_2 2^2 + a_1 2^1 + a_0 2^0$$

or $\quad a_n 2^n + \cdots + (a_3 \times 8) + (a_2 \times 4) + (a_1 \times 2) + (a_0 \times 1)$

Consider the binary number 1101_2. This number is understood to mean:

$$(1 \times 8) + (1 \times 4) + (0 \times 2) + (1 \times 1) = 8_{10} + 4_{10} + 0 + 1_{10} = 13_{10}$$

Thus, $1101_2 = 13_{10}$.

Exercise 4.5:

Convert the following binary numbers to decimal form. Answers are given below.

1. 11_2
2. 01_2
3. 10_2
4. 101_2
5. 1011_2
6. 1101_2

Section 4.6 / THE OCTAL (BASE-8) NUMBER SYSTEM

Answers

1. 3_{10}
2. 1_{10}
3. 2_{10}
4. 5_{10}
5. 11_{10}
6. 13_{10}

4.6 THE OCTAL (BASE-8) NUMBER SYSTEM

The base-8, or "octal," numbers system is just like base-10, if you are missing two fingers! Only eight digits are allowed in the octal system: 0, 1, 2, 3, 4, 5, 6, and 7.

As in the decimal and binary numbers systems, the value of an octal digit is weighted according to its position relative to other digits. The rule is:

$$a_n 8^n + \cdots + a_3 8^3 + a_2 8^2 + a_1 8^1 + a_0 8^0$$

or $a_n 8^n + \cdots + (a_3 \times 8^3) + (a_2 \times 8^2) + (a_1 \times 8^1) + (a_0 \times 8^0)$

Consider the octal number 123_8. We understand this notation to mean

$$(1 \times 64) + (2 \times 8) + (3 \times 1) = 64_{10} + 16_{10} + 3_{10} = 83_{10}$$

Exercise 4.6:

Convert the following octal numbers to decimal form. Answers are given below.

1. 213_8
2. 175_8
3. 721_8
4. 277_8
5. 671_8
6. 784_8

Answers

1. 139_{10}
2. 125_{10}
3. 465_{10}
4. 191_{10}
5. 441_{10}
6. 784 is an *illegal* octal expression! There is no "8" in the octal numbers system.

4.7 THE HEXADECIMAL (BASE-16) NUMBER SYSTEM

The hexadecimal, or base-16, numbers system is exactly like the decimal system—if you grow six extra fingers! The digits allowed in the hexadecimal numbers system are the ten from the decimal system (i.e., 0, 1, 2, 3, 4, 5, 6, 7, 8, and 9) plus A, B, C, D, E, and F. The letters A–F represent values as follows:

$$A_{16} = 10_{10} = 1010_2$$
$$B_{16} = 11_{10} = 1011_2$$
$$C_{16} = 12_{10} = 1100_2$$
$$D_{16} = 13_{10} = 1101_2$$
$$E_{16} = 14_{10} = 1110_2$$
$$F_{16} = 15_{10} = 1111_2$$

The "hex" digits are, therefore, 0, 1, 2, 3, 4, 5, 6, 7, 8, 9, A, B, C, D, E, and F. The rule for weighting in the hexadecimal numbers system is:

$$a_n 16^n + \cdots + a_3 16^3 + a_2 16^2 + a_1 16^1 + a_0 16^0$$

or $a_n 16^n + \cdots + (a_3 \times 4096) + (a_2 \times 256) + (a_1 \times 16) + (a_0 \times 1)$

Consider the hex number 123_{16}. We understand that this notation means:

$$(1 \times 256) + (2 \times 16) + (3 \times 1) = 256_{10} + 32_{10} + 3_{10} = 291_{10}$$

Exercise 4.7:

Convert the following hexadecimal numbers to equivalent decimal notation. The answers are given below.

1. $2F_{16}$
2. $3C2_{16}$
3. FF_{16}
4. $0F_{16}$
5. DE_{16}
6. $9A_{16}$

Answers

1. 47_{10}
2. 962_{10}
3. 255_{10}
4. 15_{10}
5. 222_{10}
6. 154_{10}

4.8 ARITHMETIC OPERATIONS

Binary arithmetic is used almost exclusively in computers. Although decimal arithmetic is sometimes used, it is binary that is found most often. Octal and hex numbers are used to enter data, but the CPU invariably uses the binary representation of these numbers.

Addition

The rules for binary addition are:

$$0 + 0 = 0$$
$$0 + 1 = 1$$
$$1 + 0 = 1$$
$$1 + 1 = 0 \text{ plus carry } 1$$

Example:

$$\begin{array}{ccc} 1 & 0 & 1_2 \\ +0 & 0 & 1_2 \\ \hline ? & ? & ? \end{array} \longrightarrow \begin{array}{ccc} 1 & 0 & 1 \\ 0 & 0 & 1 \\ \hline 1 & 1 & 0 \end{array}$$

Check:

$$\left. \begin{array}{l} 101_2 = 5_{10} \\ 001_2 = 1_{10} \\ 110_2 = 6_{10} \end{array} \right\} \longrightarrow \begin{array}{l} 5_{10} + 1_{10} = 6_{10}? \\ 6_{10} = 6_{10} \quad \text{CHECKS OK} \end{array}$$

Computer arithmetic units are basically *adders*. There are several digital circuits in the *arithmetic logic unit* (ALU) called *adders*. If we want to subtract, then, we must fool the computer into thinking that it is adding. We have to convert the subtrahend into a negative number (e.g., "-4_{10}" instead of "$+4_{10}$") and then add the negative subtrahend to the minuend. For example, the following are equivalent procedures:

$$\begin{array}{r} 4 \\ -2 \\ \hline +2 \end{array} \quad \text{and} \quad \begin{array}{r} 4 \\ +(-2) \\ \hline +2 \end{array}$$

In binary numbers, the negative representation is by the *two's complement* method (Section 4.9, next). To subtract in binary, therefore, we *add* the minuend and the two's complement of the subtrahend.

4.9 TWO'S COMPLEMENT NOTATION

Two's complement is a method of denoting negative numbers by using positive numbers. Let's see how this might work by considering a decimal version called 10's complement. We can see this system graphically by considering the odometer of an automobile. Let's say that the odometer reads 00000 miles. What would it read if we *back up* 10 miles? How would the odometer indicate -10_{10}? The odometer reading for -10_{10} miles would be $(00000 - 10_{10})$, or, 9990_{10}. We can, therefore, represent -10_{10} with the number 9990_{10}—a positive number.

We use a similar tactic in the binary number system. To form the two's complement of a binary number it is necessary to add one to the one's complement of the number.

Just what is the *one's complement*? The complement of any binary number is formed by making all of the zeroes into ones, and all of the ones into zeroes, e.g.:

DIGIT	COMPLEMENT
0	1
1	0

The one's complement of a multi-digit binary number is merely an expression with all of the ones converted to zero, and vice versa. The one's complement of the binary number 10110011_2 is, therefore, 01001100_2.

The two's complement is formed by adding one to the one's complement of the number.

Example:

Find the two's complement of the binary number 101101101_2.

```
Binary number         1 0 1 1 0 1 1 0 1
One's Complement      0 1 0 0 1 0 0 1 0
Add +1                                + 1
                      -------------------
Two's Complement      0 1 0 0 1 0 0 1 1
of 101101101₂
```

Section 4.11 / BINARY CODED OCTAL (BCO)　　　　　　　　　　　　39

You can subtract in a binary circuit by adding together the minuend and the two's complement of the subtrahend. For most microcomputer and minicomputer applications, there will be software instructions that will perform these subtraction operations automatically.

4.10 BINARY CODED DECIMAL (BCD)

It is frequently necessary to represent decimal, octal, or hexadecimal numbers in binary form. Anytime such numbers are used in a computer, for example, they will be stored in binary format. Probably the most common system for encoding higher base numbers into the binary number system is *binary coded decimal,* or BCD. In BCD notation, the 10 digits of the decimal numbers system are represented by equivalent four-bit binary numbers. The weighting of the BCD number follows the natural 8-4-2-1 sequence of ordinary binary numbers. The BCD numbers are:

DECIMAL	BCD
0	0000
1	0001
2	0010
3	0011
4	0100
5	0101
6	0110
7	0111
8	1000
9	1001

4.11 BINARY CODED OCTAL (BCO)

Digits of the octal number system can be represented, similar to BCD, by a three-bit binary number. For example:

OCTAL	BCO
0	000
1	001
2	010
3	011

OCTAL	BCD
4	100
5	101
6	110
7	111

We frequently see binary numbers represented by their octal "equivalents." Such a system would represent the binary number 101011101_2 in the form

$$101_2 \quad 011_2 \quad 101_2$$
$$5_8 \quad\; 3_8 \quad\; 5_8$$

So, 535_8 will sometimes be used to represent the binary number 101011101_2. This is the notation frequently used on keypads and LED displays on small computers.

A variation on the theme is the so-called *split-octal* system used on eight-bit microcomputers. The eight bits of the binary word $a_7a_6a_5a_4a_3a_2a_1a_0$ are grouped as follows:

$$a_7a_6 \quad a_5a_4a_3 \quad a_2a_1a_0$$

The most significant digit (MSD) is comprised of only two bits: a_7 and a_6. The other two octal digits are comprised of $a_5a_4a_3$ and $a_2a_1a_0$.

Example

Use split-octal notation to represent the binary number 10011001_2. The binary digits are grouped according to the method above.

$$10_2 \quad 011_2 \quad 001_2$$
$$\downarrow \quad\; \downarrow \quad\;\; \downarrow$$
$$2_8 \quad\;\; 3_8 \quad\;\; 1_8$$

So, in split-octal, 231_8 can be used as a surrogate for 10011001_2. An octal keypad can be used to input binary data into the computer. Note that the MSD in split-octal can only take on the octal digits of 0, 1, 2, or 3.

4.12 BINARY CODED HEXADECIMAL (BCH)

The BCH system is used more frequently in microcomputers than either BCO or split-octal. Most small single-board computers use hex keypads and LED hex displays to communicate between the user and the machine. The instruction sets published for such computers express the binary codes in hexadecimal form.

BCH is like BCD in that it uses a four-bit binary number. The difference between BCH and BCD is in the permissible range of the binary values. The BCD system uses only the first 10 states (0000_2 through 1001_2) of a four-bit number. The BCH, on the other hand, uses all sixteen permissible states of the four-bit binary number. The BCH system is shown below:

HEX	BCH
0	0000
1	0001
2	0010
3	0011
4	0100
5	0101
6	0110
7	0111
8	1000
9	1001
A	1010
B	1011
C	1100
D	1101
E	1110
F	1111

The BCH number DF6H, therefore, represents the binary number

$$1101_2 \quad 1111_2 \quad 0110_2 \longrightarrow \underbrace{110111110110_2}$$
$$\uparrow \qquad \uparrow \qquad \uparrow \qquad\qquad \uparrow$$
$$(D_{16}) \quad (F_{16}) \quad (6_{16}) \longrightarrow \quad DF6$$

BCH is used in microcomputers because the eight-bit data word that is typically used in those machines is easily broken into two four-bit BCH digits. For example, when you wish to enter the eight-bit word 10011101_2 into a microcomputer, we would break it into two components each represented by a single BCH digit: 9_{16} (i.e., 1001_2) and D_{16} (i.e., 1101_2). Thus,

$$10011101_2 \longrightarrow 1001_2 \quad 1101_2 \longrightarrow 9D_{16}$$

The keypad or LED display used on such simple computers will most often be either split-hex (BCH) or split-octal (BCO), with BCH being the most common.

4.13 MACHINE CODES

Some codes are used to make it easier to interface the computer to certain types of machines. Two of the simpler codes are the *excess-3 code* and the *Gray code*.

The *excess-3* code is used to perform decimal subtraction using BCD representation of the numbers. Since complement arithmetic must be used to perform subtraction operations in a computer, we find that certain disallowed states are created when BCD numbers are complemented. For example, the complement of the BCD number 0001_2 is 1110_2, which is not allowed under the rules of BCD. The solution is to use excess-3 coding. The "X-3" code for BCD is found by adding 3 to the BCD number. For example:

$$\begin{array}{lrl} \text{BCD} & 0001_2 & (\text{i.e., } 1_{10}) \\ \text{Add } 3_{10} & +0011_2 & (3_{10}) \\ \hline \text{Excess-3} & 0100_2 & (4_{10}) \end{array}$$

The 10 allowable excess-3 codes are given below:

DECIMAL	BCD	EXCESS-3
0	0000	0011
1	0001	0100
2	0010	0101
3	0011	0110
4	0100	0111
5	0101	1000
6	0110	1001

Section 4.14 / CHARACTER CODES

7	0111	1010
8	1000	1011
9	1001	1100

The Gray code scheme is used to encode certain mechanical devices such as position transducers. The advantage of Gray code is that only *one bit* of the binary number is changed each time it increments or decrements. In binary, BCH or BCD systems, going from 7_{10} to 8_{10} causes *four* bits to flip:

$$7_{10} = 0 \quad 1 \quad 1 \quad 1_2$$
$$\downarrow \quad \downarrow \quad \downarrow \quad \downarrow \quad \downarrow$$
$$8_{10} = 1 \quad 0 \quad 0 \quad 0_2$$

(All four bits changed.)

The same situation in Gray code would be:

$$7_{10} = 0 \quad 1 \quad 0 \quad 0_2$$
$$\downarrow$$
$$8_{10} = 1 \quad 1 \quad 0 \quad 0_2$$

Note that only the MSD changed in making the transition from 7_{10} to 8_{10} in excess-3 coding.

4.14 CHARACTER CODES

Several alphanumeric codes are used in computer systems to represent *characters*. When a teletypewriter, printer, or 5 × 7 dot-matrix LED display sees one of these codes it will print or display the appropriate alphabetic or numeric character.

What is a *character?* A character is a graphic symbol that represents some number, operation, or concept. We have to distinguish characters from numbers. Let's consider the case of "5." The eight-bit binary word used for the *number* "5" is 00000101_2. But the ASCII (American Standard Code for Information Interchange) code for the character "5" is 00110101_2 (35_{16}). Note well that 00000101_2 is not 00110101_2; the former is the binary number for 5_{10} while the latter is the ASCII code for the symbol "5."

5
Interfacing Techniques

5.1 OBJECTIVES

1. Interface ICs of different logic families (i.e., TTL and CMOS).
2. Interface TTL microcomputer ports to LEDs and relay drivers.
3. Interface TTL microcomputer ports with "hot" loads via optoisolators.
4. Interface the microprocessor chip with an eight-bit data bus.
5. Interface other devices with the eight-bit microcomputer data bus.

5.2 SELF-EVALUATION QUESTIONS

Before studying the material in this chapter, try answering the questions given below. These questions test your knowledge of the subject matter. If you cannot answer a particular question, then place a check mark beside it and look for the answer as you read the text.

1. Draw a typical circuit for interfacing PMOS ICs with ±12 VDC power supplies with TTL devices.
2. Draw a schematic diagram of a circuit for driving an electromechanical relay from a microcomputer output port.
3. Draw a schematic for an optoisolator interfaced with a microcomputer output port.
4. Why are bidirectional bus drivers needed for most microprocessor chips?

5.3 INTRODUCTION

The problem of interfacing microprocessor chips invariably must be faced by the instrument designer. With the *possible* exception of those which

Section 5.4 / INTERFACING DIFFERENT IC LOGIC FAMILIES

use OEM single-board computers with suitable interface connectors, the designer of the microprocessor-based instrument will need to know how to interface IC devices of different logic families, how to interface with relay drivers, how to isolate hot loads—such as the AC power mains—and how to connect both the microprocessor and other devices to the data bus.

In this chapter you will learn some of the basic designers' tricks needed to accomplish the mission of interfacing the microprocessor with other circuitry that is needed for the instrument design.

You are advised also to read the chapters on I/O and memory interfacing, as many external devices will be connected either as pseudo-memory or as an I/O port. Also, please see Chapters 19 and 20, which deal with interfacing certain other devices.

5.4 INTERFACING DIFFERENT IC LOGIC FAMILIES

Most microcomputer applications make use of TTL devices and devices from one of the metal oxide semiconductor families (NMOS, PMOS, and CMOS). In some cases, we can get away with directly interconnecting the devices, while in others some additional measures must be taken to complete the interface circuit. Figure 5-1 shows several possible solutions. Note that we are not going to deal with the obsolete families, such as RTL and DTL, or the ultrafast ECL family.

Figure 5-1(a) shows the method for interfacing PMOS devices that are operated from ±12 volt DC power supplies with TTL devices. Recall that TTL devices operate from a single +5 volt DC power supply, and are not too forgiving of other voltage levels. In Figure 5-1(a) the output of the PMOS device is connected to a pull-up resistor to the −12 volt DC power supply. A 1000 ohm current limiting resistor is connected to the input of the TTL device.

Two CMOS devices can be used to interface MOS outputs with TTL inputs, the 4049 and the 4050. These two devices are almost identical, except that the 4049 contains six inverter sections, while the 4050 contains six noninverting buffer sections. If either 4049 or 4050 devices are *operated from +5 volt power supplies*, they can then be connected directly to the input of any TTL device without further regard for interfacing. Of course, if the 4049/4050 package power supply is other than +5 volts DC, then all bets are off and some other tactic is needed. The input of the 4049/4050 can be any MOS level. An example is shown in Figure 5-1(b).

The problem of interfacing a TTL output with a CMOS input is solved in Figure 5-1(c). Recall that the TTL output is a bipolar transistor

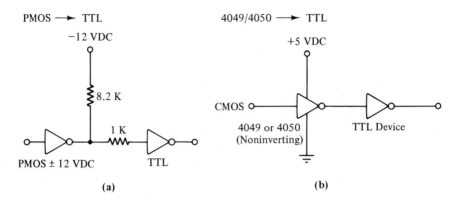
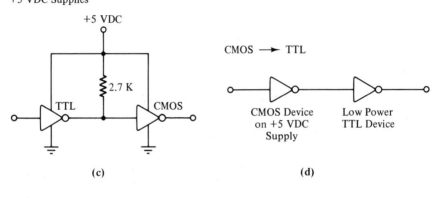
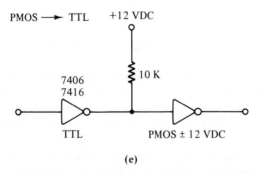

FIGURE 5-1. (a) Interfacing PMOS to TTL. (b) Interfacing CMOS to TTL. (c) Interfacing TTL to CMOS. (d) Interfacing CMOS to TTL if CMOS operated from +5 volt supply. (e) Interfacing TTL to PMOS.

Section 5.5 / INTERFACING

that acts as a current sink when in the LOW condition, while a CMOS input is a very high impedance MOSFET gate. We can connect the high impedance input directly to the output of the TTL device, as in Figure 5-1(c), provided that we supply a pull-up resistor to act as a current source for the TTL output during the LOW state (current sinks like to see current sources). A resistor in the 1 kohm to 3.9 kohm range is used for this purpose [2.7 kohm in Figure 5-1(c)].

In some cases, the CMOS type of device that is operated from a +5 volt DC power supply can be connected directly to a *low power* (74L-series) TTL device, as shown in Figure 5-1(d). This is not to be done for *regular* TTL devices.

Figure 5-1(e) shows the connection of an open-collector TTL device, such as the 7406 and 7416 inverters, to a PMOS input (when the device is operated on ±12 VDC bipolar power supplies). A 10 kohm pull-up resistor to the +12 volt DC power supply is used for the TTL output.

5.5 INTERFACING WITH LEDS, LAMPS, RELAYS, AND OTHER DEVICES.

Microcomputers used in instrumentation and process controller applications usually have an input/output (I/O) port that can be used for interfacing. The topic of creating I/O ports is treated in Chapter 10 of this book. But in this chapter, we can consider how to interface devices such as LEDs, lamps, and relays to the I/O port.

The typical microcomputer I/O port will offer eight bits. In some cases, there are separate input and output connections, while in others (such as 6502 machines that use the 6522 VIA) there are only eight bit terminals, and these are software controlled to be either input or output—but not both. An interesting feature of the 6522 device is that the I/O ports can be made inputs or outputs on a bit-by-bit basis; it is not necessary to specify an entire port as either I or O.

Figure 5-2(a) shows the basic connection to the output port for all of the circuits shown in this section. The input of either an inverter or a non-inverting buffer is connected directly to the output port bit—bit B0 is shown in this case, but any bit could be used. We could connect *some* loads directly to the output port pin, but that is usually unadvisable because of the limited current capability of the terminal. Typically, only one or two TTL loads can be driven directly by the microcomputer port.

The typical light emitting diode (LED) requires from 5 to 100 mA of current in order to light to a reasonable brilliance. The microcomputer output port, however, cannot handle such a current. Figure 5-2(b) shows the method for connecting the inverter of Figure 5-2(a) to an LED. Use an

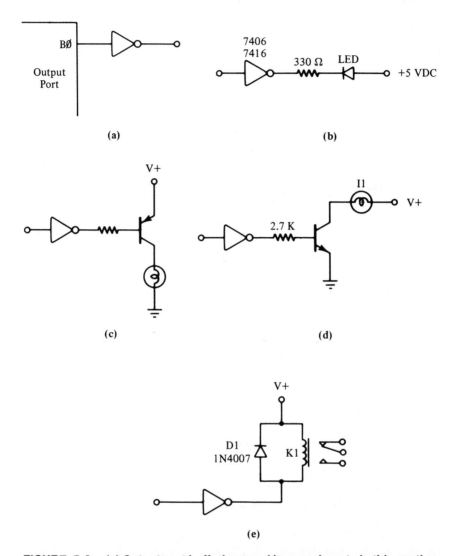

FIGURE 5-2. (a) Output port buffering used in experiments in this section. (b) Driving an LED indicator from a computer output port. (c) Lamp driver (PNP). (d) Lamp driver (NPN). (e) Relay driver.

inverter that is advertised as an *open-collector* TTL device. In such a TTL IC, the return path to the V+ is not made internally. The transistor in the output stage has its collector connected directly to the output terminal of the inverter. This transistor can withstand potentials to either 16 volts DC or 30 volts DC, depending upon type. In Figure 5-2(b) the LED

Section 5.5 / INTERFACING

is shown connected to the +5 volt power supply, but that is not strictly necessary—any positive polarity supply will do. The 330 ohm resistor is used to limit the LED current to a safe value. The value selected here is for most common LEDs, and with a +5 volt power supply, limits the current to 15 mA.

Interfacing to incandescent lamps is shown in Figures 5-2(c) and 5-2(d). Most lamps require currents greater than typical LEDs, but they remain in use by many designers. Typical lamps require from 40 mA to 1 ampere. Because of the high current, it is often necessary to use an external transistor to control the lamp (the transistor acts as a switch), while the inverter/buffer will control the transistor base junction. In Figure 5-2(c) we see the use of a PNP transistor to control the lamp. The PNP transistor will conduct when its base is more negative (or less positive) than the emitter. The lamp is connected between the collector and ground, so it will not illuminate until the transistor conducts. As long as the output of the inverter is HIGH, indicating a LOW on the microcomputer output port terminal, the base and emitter of the PNP transistor are at the same potential; the transistor is turned off so the lamp will not light up. But when the output port bit becomes HIGH, the output terminal of the inverter drops LOW, placing the base of the PNP transistor at a potential close to ground. This will forward bias the transistor, and cause it to conduct. The lamp will now turn on.

A lamp driver made from an NPN transistor is shown in Figure 5-2(d). This circuit works in the opposite sense from Figure 5-2(c). In this case, the base of the transistor must be made more positive than the emitter in order to turn on the transistor. Placing a HIGH on the output port bit will cause the inverter output to be LOW, so the transistor is cut off; the lamp will not turn on. However, when a LOW appears on the output bit, the *b-e* junction of the NPN lamp driver transistor becomes forward biased—the lamp turns on. The resistor in series with the base is used to limit the current to a safe value, yet it must not have a value that is high enough to cause insufficient bias on the transistor.

A relay driver circuit is shown in Figure 5-2(e). Relays are sometimes used to handle either very high current loads or "hot" loads such as the 110-volt AC power mains. An open-collector inverter is used, with the relay coil forming the pull-up between the output of the open-collector TTL inverter and the V+ power supply. If appropriate inverters are used, then up to 28-volt DC relays can be accommodated. The diode in parallel with the relay coil is absolutely essential. It functions to dampen the high voltage spike caused by "inductive kick" when the relay coil is de-energized. This spike can easily reach several hundred volts, and may reach over 1000 volts in some cases. Such a spike would burn out the inverter and other semiconductors in the circuit.

5.6 TRI-STATE DEVICES

Most microprocessors use a common data bus for many operations. Data to and from the CPU flow across this bus to and from memory locations, I/O ports, and so forth. If any one device is LOW, then that bit of the data bus is held permanently LOW—an unsatisfactory state of affairs. We can, however, bus together many different devices that share a common line by using tri-state outputs. Normal TTL and CMOS devices offer only two output states, HIGH and LOW. In the HIGH condition, they form a low resistance to the V+ power supply, while in the LOW condition they form a low resistance to ground. The mechanism for representing these conditions is shown in Figure 5-3 as switch $S1$. When the input terminal is LOW, then $S1$ is connected to the low resistance to ground, but when the input terminal is HIGH, $S1$ is connected through a low resistance to +5 VDC. The tri-state device offers a third state in which the output terminal is *disconnected* from the output terminal. This situation is represented by S2 being open-circuited. In point of fact, the output terminal is not totally disconnected from $S1$, but is at a high impedance—very high. A *chip enable* terminal will either disconnect the output or connect it, depending upon whether the CE is HIGH or LOW.

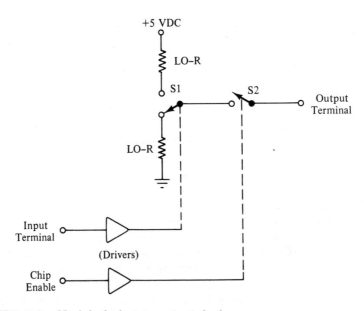

FIGURE 5-3. Model of tri-state output devices.

5.7 BUS DRIVERS

Several IC devices are used to drive the address and data buses in microcomputers. The data output terminals of most microprocessor chips have only a limited amount of drive capability, typically around 3 mA—the amount of current needed to drive only two TTL input loads. In order to

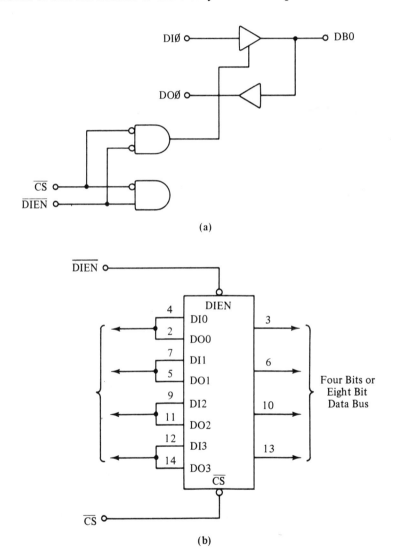

FIGURE 5-4. (a) Bidirectional bus driver (one section shown). (b) 8212 bidirectional bus driver.

drive the large number of devices that might be connected to the data bus, we must use a high power bus driver between the microprocessor and the bus. Furthermore, this bus driver must be *bidirectional* because data flows into and out of the microprocessor along the same bus.

Figure 5-4(a) shows the basic method used to accomplish bidirectionality in the Intel 8216 and 8226 (inverting and noninverting versions of the same theme) devices. The data bus output terminal (DB0) is connected to the output of one buffer and the input of another. Separate data input (DI0) and data output (DO0) terminals are provided. Two control terminals are used in this chip, $\overline{\text{DIEN}}$ and $\overline{\text{CE}}$. The $\overline{\text{DIEN}}$ is the data direction enable, while the $\overline{\text{CE}}$ is the chip enable.

Figure 5-4(b) shows the connection of the four-bit 8216 and 8226 devices to four bits of the data bus. For an eight-bit data bus, two similarly connected devices are needed. In this case, the DB0 through DB3 terminals are connected to the lines of the data bus, while the respective DI0-DO0 and DI0-DI3 lines are jumpered together (4-2, 7-5, 9-11, and 12-14) to connect to the data bus output lines of the microprocessor chip. The $\overline{\text{DIEN}}$ and $\overline{\text{CE}}$ terminals will be connected to the appropriate control lines. These are not shown here because they are different for each microprocessor.

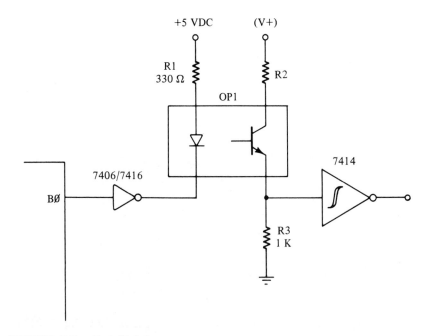

FIGURE 5-5. Optoisolator.

If a unidirectional bus driver is needed, then use devices such as the four-bit 74125 and 74126 devices or the eight-bit 74LS244/74LS245 devices. These are tri-state devices, but not bidirectional.

5.8 USING OPTOISOLATORS

There are times when it is undesirable to drive a load directly. Some of these are "hot" loads such as the AC power line, or, higher than usual DC potentials. The optoisolator consists of an LED on the "input" side, and a phototransistor on the output side. The LED is connected in the same manner as before. A HIGH on the B0 output of the I/O port will turn on the LED, causing it to illuminate the phototransistor. The phototransistor will now conduct, causing a voltage drop across resistor $R3$. This voltage can be used to drive the gate of a triac or a circuit such as the Schmitt trigger, as shown in Figure 5-5.

6
Microprocessor Support Chips

6-1 OBJECTIVES

1. State the different types of support chip available to each microprocessor family.
2. Design applications around these support chips.

6-2 SELF-EVALUATION QUESTIONS

Before studying the material in this chapter, try answering the questions given below. These questions test your knowledge of the subject matter. If you cannot answer any particular question, then place a check mark beside it and look for the answer as you read the text.

1. What is the *byte directional mode* in the Z80-PIO?
2. The Z80-SIO is a _____ I/O port.
3. "DMA" stands for _____ _____.
4. What are three modes of operation for the Z80-DMA device?

6.3 SUPPORT CHIPS

In order to make the typical microprocessor chip think it is a real live computer we need additional, external circuitry. In some commercial products, this external circuitry takes the form of TTL and/or CMOS devices connected to perform the desired function. But some companies make it easier to make a computer by using certain external special-function integrated circuits.

Section 6.4 / Z80-PIO   55

Two special-function devices are used to provide serial and parallel input/output capability. The Z80-SIO device is a serial I/O chip, while the Z80-PIO is a parallel I/O port. These devices are second-sourced by Mostek under the type numbers MK3884 (Z80-SIO) and MK3381 (Z80-PIO).

There is also a direct memory access device called the Z80-DMA (Mostek MK3883). Direct memory access in a computer allows the external memory to be written to, or read from, by a peripheral device without first going through the CPU. This allows the operation to be performed much more rapidly and is conservative of CPU time—a precious commodity in some applications.

The Z80-CTC (Mostek MK3882) is a four-channel, multimode counter/timer circuit. It provides counter and timer capability in Z80-based microcomputer systems.

6.4 Z80-PIO

The Zilog Z80-PIO (Mostek MK3881) is used as a parallel I/O port controller. It contains two ports and is user programmable. The Z80-PIO contains two completely independent, eight-bit bidirectional ports. Complete handshaking capability is permitted, so the device can be used for synchronous transfers.

The Z80-PIO can be programmed to operate in four different modes: *byte output, byte input, byte bidirectional bus* (port A only), and *bit control*.

The byte output mode, also called mode-0, is used to allow the CPU to write data to the peripheral via the CPU data bus. If mode-0 is selected, a *data write* operation causes a handshake signal (*ready*) to be generated. This signal is used to let the peripheral know that the data are available and valid. Note that the data remain available, and the *ready* signal remains HIGH, until a strobe is received back from the peripheral.

The *byte input mode,* also called mode-1, allows the selected port to behave as an input port only. When a *data read* operation is performed by the CPU, the PIO will issue a *ready* signal to the peripheral. This tells the peripheral that the Z80 CPU is now in a condition to receive the input data. The peripheral responds by issuing a strobe that causes the data to be transferred to the data input register of the PIO.

The *byte bidirectional mode,* also called mode-2, uses the port as a bidirectional, eight-bit I/O port. Mode-2 uses all four possible handshake lines. Because of this restriction, only port-A can be used in the bidirectional mode.

The *bit control mode,* also called mode-3, is used for status and control applications. Mode-3 does not make use of the handshake signals. This mode is used to define which port data bus lines will be inputs and which will be outputs. The next word fed to the PIO after mode-3 is selected must define these conditions.

Figure 6-1 shows the pinouts for the Z80-PIO, while below are the definitions of the different types of pins.

D0-D7   These pins connect to the Z80 CPU data bus, and are both bidirectional and tri-state. All command signals and data passed between the CPU and the PIO, in either direction, must be passed over these lines.

B/A SEL   This active-HIGH input will select either port A or port B. A LOW on B/A SEL will select port A, whereas a HIGH will select port A.

C/D SEL   This active-HIGH input selects the type of data transfer to take place between the CPU and PIO. A LOW on this line tells the PIO that the data on the Z80 data bus are I/O data. But a HIGH will tell the PIO that the data being transferred are a command for the port selected by B/A SEL.

$\overline{CE}$ Active-LOW input that acts as a chip enable. A LOW on this terminal allows the PIO to accept command/data inputs from the Z80 CPU during any write cycle, or to send data to the Z80 CPU during any read cycle.

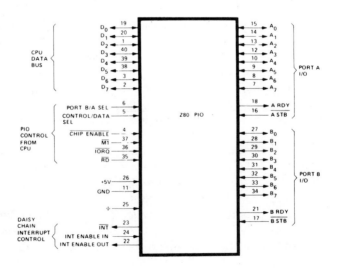

FIGURE 6-1. Z80-PIO pinouts.

$\overline{\text{M1}}$	This terminal synchronizes the PIO to the CPU, and is generally connected to the similarly named terminal on the CPU chip. Indicates that an M1 machine cycle is in progress.
$\overline{\text{IORQ}}$	Input/output request line from the Z80 CPU chip that is part of the sync system. Usually connected to the similarly named terminal on the Z80 device.
$\overline{\text{RD}}$	Active-low input that detects the read cycle of the Z-80.
IEI	Interrupt Enable Input. This is an active-HIGH input.
A0-A7	Tri-state, bidirectional address bus for port-A.
$\overline{\text{A STB}}$	Active-LOW input that strobes port-A from peripheral device.
A RDY	Active-HIGH output signals that the A-register is ready.
B0-B7	Tri-state, bidirectional, address bus for port-B.
$\overline{\text{B STB}}$	Active-LOW input that allows peripheral device to strobe port-B.
B RDY	Active-HIGH output that signals that the B-register is ready.

6.5 Z80-SIO

The Z80-SIO device is a serial I/O chip that interfaces directly with the Z80 CPU chip. It is similar to the Z80-PIO in that it is a programmable two-channel device. The SIO, however, transmits the data in the *serial* stream, i.e., one bit at a time. Parallel transfer is, of course, faster in most cases. But often a serial transfer is preferred because it reduces the hardware overhead between the computer and the peripheral with which it is communicating. Even when the "run" is only a short distance, it is often much less costly to use a serial data transfer because only one pair of wires, one telephone line, or one radio communications channel is required. The Z80-SIO is designed to handle just about any reasonable serial bit protocol. Like the other chips of the Z80 family, it is operated from a single +5-volt DC supply and uses only a single-phase clock.

The two channels (also labeled A and B, as in the PIO device) are totally independent of each other, except for power supply and CPU bus connections. The SIO channels are full-duplex, so data can be transmitted and received simultaneously. The Z80-SIO allows data rates from zero to 550,000 bits per second.

Both receiver and transmitter registers are fully buffered. But in the

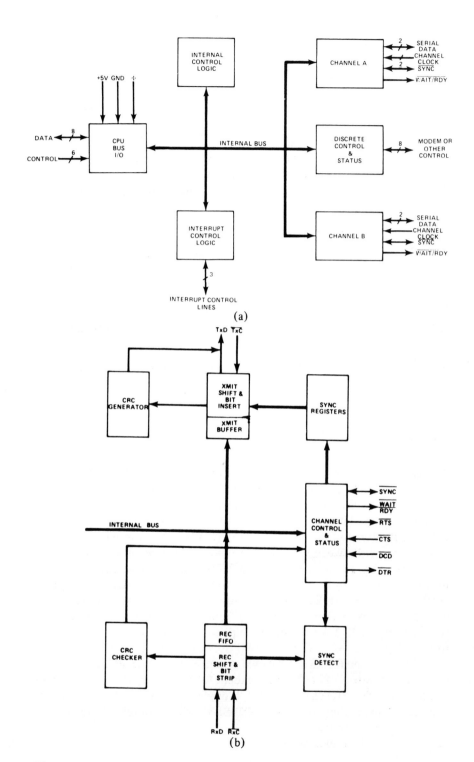

Section 6.5 / Z80-SIO

case of the transmitter section, the registers are doubly buffered. The receiver registers, on the other hand, are quadruply buffered.

The Z80-SIO is capable of *asynchronous* operation (in which it behaves much like an ordinary UART, but with a Z80-system flavor), *synchronous binary* operation, and HDLC/IBM-SDLC operation. The SIO provides eight MODEM control inputs/outputs, allows daisy chain priority interrupt logic to automatically provide the vector word, and permits both CRC-16 and CRC-CCIT $(-0/-1)$.

The SIO looks very much like the ordinary UART in its asynchronous mode. It can be programmed for 5, 6, 7, or 8 eight-bit words. Like the UART, it will provide 1, 1.5, or 2 stop bits at the end of each transmitted word. The CPU, incidentally, need not provide these bits; the SIO adds them to the word received from the CPU before the word is transmitted. Also, like the UART, the SIO will provide parity bits (even, odd, none), and detection of parity, framing errors, and overrun. Unlike most UARTs, however, the SIO also provides for the generation and detection of breaks. Clock rates of 1X, 16X, 32X, and 64X the data rate are permitted.

Figure 6-2 shows the organization of the Z80-PIO device. In Figure 6-2(a) we see the overall block diagram of the device, while Figure 6-2(b) shows the block diagram for the channels. The input section from the CPU receives eight data bus lines and six control signal lines. Once inside, the device operates from an internal bus not accessible to the outside world. There are two sections for channels A and B, some internal control logic, the interrupt section, and a discrete control section (used with MODEMs and other controlling devices).

The pinouts for the Z80-SIO are shown in Figure 6-3, and are listed in detail below:

D0-D7	Tri-state, bidirectional data bus to/from Z80-CPU and rest of Z80 system.
B/A	Channel A/B select. Channel A is selected when this pin is LOW, and channel B is selected when it is HIGH.
C/D	Control/data select. If this input is HIGH, then the control mode is selected, but if it is LOW, then the data mode is selected.
$\overline{M1}$	Active-LOW input that detects the M1 machine cycle in Z80.

FIGURE 6-2. Z80-PIO organization. (a) Overall internal block diagram; (b) channel block diagram.

MICROPROCESSOR SUPPORT CHIPS / Chapter 6

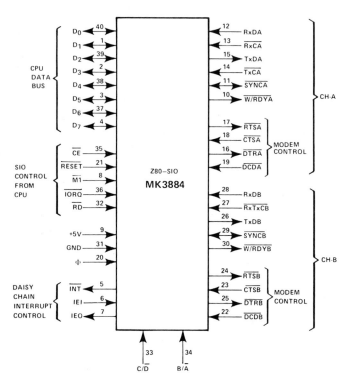

FIGURE 6-3. Z80-SIO pinouts.

$\overline{\text{IORQ}}$ — Active-LOW input that detects the Input/Output ReQuest state of the Z80 CPU.

$\overline{\text{RD}}$ — Active-LOW input that detects the read cycle of the Z80 CPU.

Φ — Clock terminal.

$\overline{\text{RESET}}$ — Active-LOW input that resets the system. Placing a LOW on this terminal has the following results: both receivers and transmitters are disabled, TDA/TDB are forced marking, modem controls are forced HIGH, and all interrupts are disabled. *Note:* The control registers of the SIO *must* be rewritten from the CPU before the SIO can be again used.

IEI — Active-HIGH interrupt enable input.

IEO — Active-HIGH output. Note that IEI/IEO are used together to form a daisy-chain priority interrupt control function.

INT	Active-LOW output to the interrupt request line of the Z80. Note that this terminal is an open-drain type.
WAIT READY A WAIT/ READY B	These lines, one of each channel, have two principal functions. In one case, they can be used as ready lines for the Z80 DMA (direct memory access) controller. In another, they can be used to synchronize the Z80 CPU to the Z80-SIO (i.e., to sync the data rate between CPU and SIO).
CTSA CTSB	These lines, one for each channel, provide a *clear to send* function. Both are active-LOW inputs. If programmed for *auto enable,* then these pins will act as transmitter enable controls. But when not programmed for auto enable, they can be programmed for general control purposes. *Note:* These pins are buffered through Schmitt-trigger circuits, thereby allowing slow rise-time signals.
DCDA DCDB	Data Carrier Detect. These two active-LOW inputs serve as receiver enable control signals.
RDA/RDB	Active-HIGH receiver data inputs.
TDA/TDB	Active-HIGH transmit data outputs.
RCA/RCB	Schmitt-trigger buffered, active-LOW receiver clock inputs.
TCA/TCB	Same as above, but transmitter clocks.
RTSA/ RTSB	Active-LOW outputs providing *request-to-send* signals.
DTRA/ DTRB	Active-LOW outputs providing *data-terminal-ready* signals.
SYNCA/ SYNCB	Used for synchronization of external characters.

6-6 Z80-DMA

The Z80-DMA (Mostek MK3883) is a *direct memory access* controller. This type of operation is very useful in a computer. It speeds up direct transfers between an external device, or peripheral, and the memory because it allows bypassing of the CPU. Ordinarily, if you wanted to transfer a data word from some peripheral device and a specific memory location, you would have to execute an input instruction to move the data into the accumulator first. Then a second instruction would be required in order to move the data from the CPU to the desired memory location. Unless the data are to be used immediately after input, this would be a waste of valuable time. DMA allows the data to be placed directly into the desired location from the peripheral.

The DMA chip allows three modes, or classes, of operation: *transfer only*, *search only*, and *search-transfer*. There are also four types of operation: *single byte at a time*, *continuous burst* (as long as ports are ready), *continuous* (CPU locked out), and *transparent* (i.e., it steals time from refresh cycles).

Three types of interrupt are allowed. In one case, the DMA chip will interrupt the CPU only when a match to a desired word is found. It will also interrupt on *end-of-block* or *ready*. The DMA can be enabled, disabled, or reset totally under software control.

Figure 6-4(a) shows the pinouts for the Z80-DMA, while Figure 6-4(b) shows the internal block diagram. The pinout functions are discussed below:

A0-A15	System address bus (from Z80 and memory). This 16-bit address bus can, like the Z80 bus, address all 64K of allowed memory.
D0-D7	Data bus from CPU and memory. These tri-state input/output pins carry three types of data: commands from the Z80 CPU, DMA status (from memory/peripherals), and data from the memory/peripherals.
Φ	System clock.
$\overline{M1}$	Active-LOW input detects the M1 machine cycle in the Z80 CPU.
$\overline{IORQ}$	Used as an input/output request to/from the CPU bus.
$\overline{MREQ}$	Used as a memory request to/from Z80 system bus.
$\overline{RD}$	Read to/from Z80 CPU bus.
$\overline{WR}$	Write signal to/from Z80 CPU bus.
$\overline{CE/WAIT}$	May be used as either chip enable or $\overline{wait}$.
$\overline{BUSRQ}$	Bus request is used to request control of the data bus from the Z80 CPU.
$\overline{BAI}$	Input that tells the Z80-DMA that the CPU has granted it control of the bus. it is a bus acknowledge input.
$\overline{BAO}$	Bus acknowledge output that allows daisy chain connection of DMA-requesting peripherals.
$\overline{INT}$	Active-LOW output that tells the Z80-CPU that an interrupt is requested.

Section 6.6 / Z80-DMA

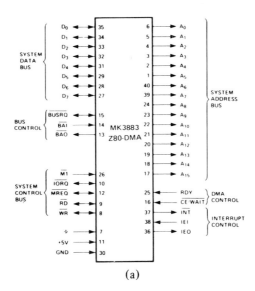

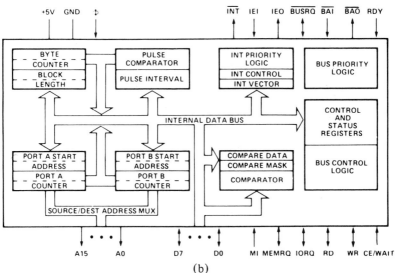

FIGURE 6-4. (a) Z80-DMA pinouts; (b) Z80-DMA block diagram.

IEI	Active-HIGH interrupt enable input.
IEO	Active-HIGH interrupt enable output. Forms ability to daisy chain, when used in conjunction with IEI.
RDY	Active-HIGH/LOW (i.e., programmable) input that tells the Z80-DMA when a peripheral device is ready for a write/read operation.

6-7 Z80-CTC

The Z80-CTC (Mostek MK3882) is a universal counter-timer chip that can provide all of the counter/timer requirements for a Z80-based computer. There are four independent channels in the Z80-CTC. Consistent with the design of the rest of the Z80-family, this device requires only a single +5-volt dc power supply and a single-phase clock. Each of the four channels can operate as either a counter or a timer.

The Z80-CTC pinouts are shown in Figure 6-5, and their respective descriptions are given below:

D0-D7   Bidirectional tri-state data bus to/from CPU.

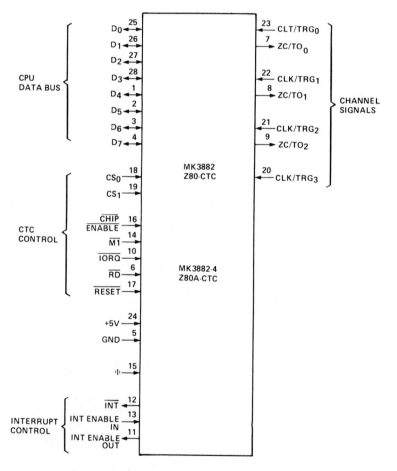

FIGURE 6-5. Z80-CTC pinouts.

CS0-CS1 Active-HIGH channel select inputs.
$\overline{\text{CE}}$ Active-LOW chip enable input.
Φ System clock.
$\overline{\text{M1}}$ Active-LOW input from CPU that detects the M1 machine cycle.
$\overline{\text{IORQ}}$ Active-LOW input that detects the input/output request state of the CPU.
$\overline{\text{RD}}$ Active-LOW input that detects the Z80-CPU read cycle.
IEI Active-HIGH interrupt enable input.
IEO Active-HIGH interrupt enable output. Used with IEI to permit daisy chaining.
$\overline{\text{INT}}$ Active-LOW, open-drain, output to the Z80-CPU interrupt request input.
$\overline{\text{RESET}}$ Active-LOW, reset input.

7
Solving Problems with the Microprocessor

The microprocessor has been touted as *the* major breakthrough for engineers who design electronic instrumentation and industrial process controllers. A few years ago, professors at major engineering schools were telling their students to develop skills in microcomputer/microprocessor technology. It was said at that time that "engineers without these skills will be *unemployable* within 10 years." At the time, that claim sounded like little more than hype to make students want to sign up for a new course. But as time went on, we found that it was actually conservative—five years was nearer the truth.

What makes the microprocessor superior to other methods? After all, isn't it just another integrated circuit? Yes, but only in the way a 500-lb. Bengal tiger is "only another cat." The microprocessor is a full-fledged programmable digital computer. In some versions (e.g., the Z80 or 6502) the chip is a microprocessor, while others (e.g., the 8048) are complete microcomputers-on-a-chip. Microprocessors become microcomputers by adding external memory and I/O chips.

Most engineers and engineering students are familiar with the business of solving instrumentation problems and control problems using analog circuits, ordinary digital logic circuits, and devices such as switches and relays. But microprocessor solutions are a little different. There are two main goals in microprocessor-based instrumentation:

1. Replacement of digital logic devices with software.
2. Replacement of some *analog* circuit functions with software.

Before we hear too many howls of anguished protest over the second point, let me hasten to point out—and dispense with—a myth about digital instrumentation that is often held by analog designers. It is often claimed that "digital" contains a built-in error, i.e., the *discrete quantiza-*

tion error. This error is inherent in digital implementation of circuits. True, by digitizing analog signals we do inherit a certain basal error rate. But the argument that analog circuitry is more accurate because it permits an infinite number of discrete values between the lower and upper limits of a voltage or current range is not supportable. Analog circuits *also* contain substantial errors: amplifier gain error, offset voltage error, drift, and readout/display resolution. The digital implementation of a circuit function often produces superior accuracy because it eliminates the analog errors, and digitization error is low. Incidentally, digitization error is reduced by a factor of two for every bit that is added to the length of the data word.

Another myth is the notion that, by using a computer, all of our problems are solved. It is a common fallacy that computers *solve problems;* they DO NOT. *Engineers solve problems. Programmers* solve problems. The computer can no more "solve problems" than a saw or hammer can build a wooden house. In designing electronic instruments using a microcomputer or microprocessor technique, *you must first solve the problem at least at the flow chart level.* Only after you have produced a step-by-step plan of action (i.e., an *algorithm*) that will do the job can you begin to select the components that will be used, architectures, and software. Just as the analog circuit designer begins with a block diagram solution of the problem *on paper,* the microcomputer-oriented designer must begin with an algorithm on paper and, sometimes, a system block diagram.

We can see this process by considering a trivial example: a simple traffic light controller. Let's assume that the design specifications require the following:

1. East-west green for 30 seconds, while north-south is red.
2. E-W is yellow for 7 seconds, while N-S is green.
3. E-W is red, and N-S is green, for 30 seconds.
4. E-W is red, and N-S is yellow for 7 seconds.
5. Repeat steps 1-4, in sequence, indefinitely.

Figure 7-1 shows a proposed solution to this problem. We write a program that will cause the various on-off states for the lights to occur at appropriate times, and for appropriate durations.

Next, we must allocate our resources. Suppose that we are going to use a 6502-based single-board "controller," with a 6522 PIA for I/O interfacing. This arrangement will allow us to use the Rockwell AIM-65 computer as a development system. This approach is especially valid if we use the same memory locations for the 6522 PIA as in the AIM-65, i.e.:

LOCATION	FUNCTION		
A000	Port B Output Data Register (ORB)		
A001	Port A Output Data Register (ORA) Controls handshake		
A002	Port B Data Direction Register (DDRB) } 0 = Input		
A003	Port A Data Direction Register (DDRA) } 1 = Output		
	Timer	$R/\overline{W} = L$	$R/\overline{W} = H$
A004	T1	Write T1L-L	Read T1C-L Clear T1 Interrupt Flag
A005	T1	Write T1L-H & T1C-H T1L-L → T1C-L Clear T1 Interrupt Flag	Read T1C-H
A006	T1	Write T1L-L	Read T1L-L
A007	T1	Write T1L-H Clear T1 Interrupt Flag	Read T1L-H
A008	T2	Write T2L-L	Read T2C-L Clear T2 Interrupt Flag
A009	T2	Write T2C-H T2L-L → T2C-L Clear T2 Interrupt Flag	Read T2C-H
A00A	Shift Register (SR)		
A00B	Auxiliary Control Register (ACR)		
A00C	Peripheral Control Register (PCR)		
A00D	Interrupt Flag Register (FR)		
A00E	Interrupt Enable Register (IER)		
A00F	Port A Output Data Register (ORA) *No effect on handshake*		

For those who are unfamiliar with the 6522 and the AIM-65, let us discuss the I/O protocol. The 6522 (which is used for I/O in the AIM-65) has two I/O ports, designated A and B. These ports are treated as memory locations because the 6502 microprocessor uses memory-mapped I/O. In the AIM-65, these ports are assigned to memory locations $A000_{16}$ (port B) and $A001_{16}$ (port A).

The I/O registers in the 6502 are bidirectional, meaning that they may be used as either input or output. The direction is controllable on a bit-by-bit basis (e.g., BA0 can be an input, while BA1 is an output). The direction of a bit in either port is set by the bit applied to the corresponding position in a *Data Direction Register* (DDRA and DDRB). A HIGH written to a DDR bit will make the corresponding bit in the I/O port an output; a LOW in the DDR bit will make the port bit an input. In

SOLVING PROBLEMS WITH THE MICROPROCESSOR / Chapter 7

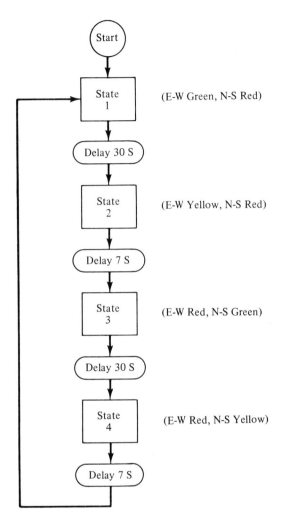

FIGURE 7-1. Traffic light flow chart.

the AIM-65, DDRA is located at $A003_{16}$, while DDRB is located at $A002_{16}$. By writing 00 to A002, we will make port B an input. If we had written FF_{16} to $A002_{16}$, however, we would make all of port B an output port.

Now back to our traffic light controller. Figure 7-2 shows a circuit to connect the lights (which, in this example, are represented by LEDs) to an AIM-65 microcomputer. Each LED is driven by an open-collector TTL inverter (one section of a 7405 or 7406 hex inverter). When the input

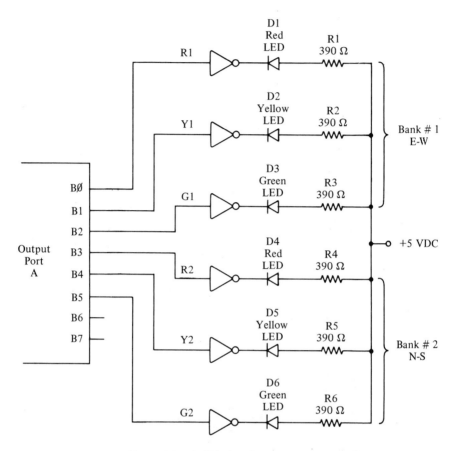

Note: Lamp is ON when Output Port Bit is HIGH

FIGURE 7-2. Traffic light interface.

of the inverter is HIGH, then its output is LOW, which turns on the LED by completing its ground path. A HIGH applied to the corresponding port A output bit will turn on the LED. For example, writing 00100000 to port A will make bit B5 HIGH, turning on LED D6 (north-south green light).

The idea is to write a program for the AIM-65 that will place a HIGH on those output bits in port A that connect to the lights that must be turned on, and a LOW to all other bits. For example, state 1 requires R2 and G1 to be on, and all others off. The binary word for this state is 00001100_2 or $0C_{16}$ in hexadecimal (see Figure 7-3). This state is held for 30 seconds, and then state 2 is generated.

Section 7.1 / PROGRAMMING EXERCISES 71

Lamps Bits			G2	Y2	R2	G1	Y1	R1	State Code at Output	
	B7	B6	B5	B4	B3	B2	B1	B0		
State									Binary Word	Hex
1	X	X	L	L	H	H	L	L	00001100	0C
2	X	X	L	L	H	L	H	L	00001010	0A
3	X	X	H	L	L	L	L	H	00100001	21
4	X	X	L	H	L	L	L	H	00010001	11

N-S ← → ← E-W →

X = Don't Care (Assign Logical-0 for Convenience)
H = High (i.e., Logical-1)
L = Low (i.e., Logical-0)

FIGURE 7-3. Traffic light state table.

7.1 PROGRAMMING EXERCISES

Build the circuit of Figure 7-2 for these exercises. Use either *Vector* DIP perfboard, an IC breadboarding "socket" or a "digital breadboard" such as the AP Products, Inc. *Powerace 102*.

    7-1. Write a program that will cause the N-S direction to flash red, and E-W to flash yellow, on one second intervals. (a) Alternate between E-W and N-S, (b) flash E-W and N-S simultaneously. Submit the flowchart and program listing to your instructor.

    7-2. Write a program that will implement the algorithm of Figure 7-1. Submit your flowchart and program to the instructor.

    7-3. Write a program that will meet the following specifications:
        1. Permit E-W traffic flow all the time *unless* N-S traffic appears (see Figure 7-4 for pedestrian and auto sensors).
        2. If pedestrian traffic appears, switch lights to permit N-S traffic immediately.
        3. If one car appears in the N-S lane, switch to permit N-S traffic after 30 seconds. If two cars appear in the N-S lane, wait no more than 15 seconds (including the time elapsed since the first car arrived), and if three cars appear, switch immediately.

    7-4. Write a program that will implement 7-1 through 7-3 above, depending upon the setting of switch S1 in Figure 7-5.

72 SOLVING PROBLEMS WITH THE MICROPROCESSOR / Chapter 7

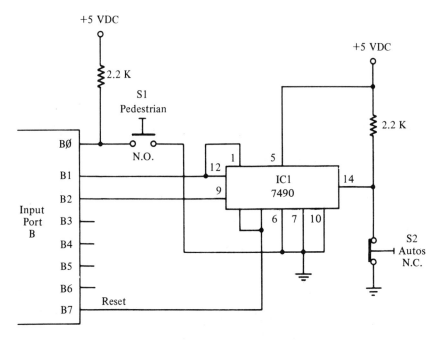

(Alternatively, Use the PIA Internal Timer Instead of IC1)

FIGURE 7-4. Pedestrian and autos sensor mock-up.

7.2 INTEGRATION AND DIFFERENTIATION

The mathematical processes of integration and differentiation are fundamental to the operation of many instruments. In analog designs, these processes were carried out using appropriate operational amplifier circuits. In digital instruments, however, we must use certain numerical methods of integration and differentiation. We must also be aware of certain anomalies in the data that would ruin the process (these will be discussed later). In this section we will consider the more elementary numerical methods, with the understanding that the wise student will either take a course, or read a textbook, on numerical methods of computation.

The process of differentiation is used to find the instantaneous rate of change of a function. In the computerized instrument, a voltage input function is converted to a binary word by an analog-to-digital converter, and it is on the binary data that the machine operates.

Figure 7-6 shows a crude method of finding the derivative of an input signal. We can use the natural clock cycle time of the computer for the $T_2 - T_1$ data or use a software timer to create longer periods. The A/D

Section 7.2 / INTEGRATION AND DIFFERENTIATION

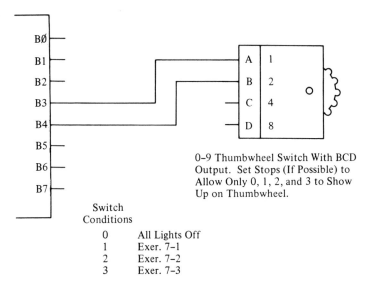

FIGURE 7-5. Thumbwheel switch connections (see text).

converter will take two samples of the input signal, Y_1 and Y_2. The derivative is approximately

$$\frac{\Delta Y}{\Delta T} \approx \frac{Y_2 - Y_1}{T_2 - T_1}$$

provided that ΔT is not too long.

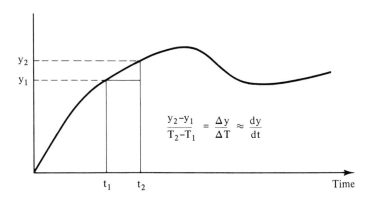

FIGURE 7-6. Differentiation by numerical method.

To make a continuous derivative output, take continuous samples of the signal $(Y_n, \ldots, Y_3, Y_2, Y_1)$ and make the calculations $(Y_2 - Y_1)$, $(Y_3 - Y_2)$, $\ldots (T_n - T_{n-1})$. The time factor ΔT will be the same for each and cannot be faster than the cycle time of the A/D converter.

If an application requires an analog output that is proportional to the derivative of the analog input signal, then output the results of the calculation to a digital to analog converter (DAC). This circuit will produce an analog voltage that is proportional to the derivative of the analog input.

One problem faced by researchers in chemistry and the life sciences is the long amounts of time required for some of the processes to take place. Changes in physiological (living) systems, for example, often takes seconds, or even minutes. Electrical engineers used to faster signals may not be aware that the slow signal is difficult to differentiate using analog circuitry. In fact, it is often impossible. Active differentiators require operational amplifiers to work. The input bias currents will eventually charge the input capacitor used for differentiation, causing the operational amplifier to latch up. Long before the amplifier latches up, however, the offset bias current will create a substantial artifact in the data. The microprocessor, however, can be used to make a differentiator for slow signals. An A/D converter for the uP input, and a DAC for the output, make the instrument look like an analog device.

Integration can also be performed using a microprocessor. The "computer" version is a bit more accurate than the typical analog intergrator that uses an operational amplifier. The problem with the analog circuit is that output bias voltages will tend to charge the feedback capacitor without regard for the input signal. Figures 7-7(a) and (b) show two methods for making digital integrators.

Method I [Figure 7-7(a)] is the more crude, but it is sufficient for many applications. The A/D converter will input a series of voltage values to the computer: V1, V2, V3, V4, V5, etc. This will divide the area under the waveform into a series of individual rectangles. The sum of these rectangles' area is the integral of the waveform over the time of measurement. If the time periods are equal for all samples, then we can multiply each input value (V) by ΔT. Then, by summing these values in a holding register somewhere, we will have the integral.

But there can be significant error in this measurement if the time period is too great. Very often, we find that some factor makes it difficult to make the time period for each rectangle sufficiently small to permit accurate integration. The shaded area of Figure 7-7(a) shows the amount of error.

A more accurate method of integration is shown in Figure 7-7(b). We include those errant shaded areas by using a slightly more complex formula for making the integration. In the case of Figure 7-7(b) we make the

Section 7.2 / INTEGRATION AND DIFFERENTIATION

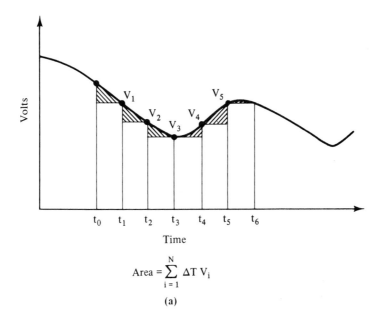

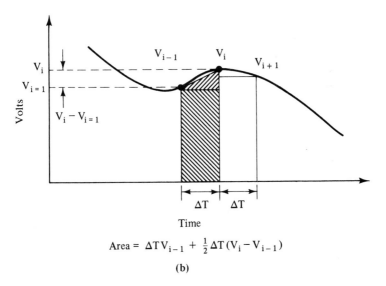

FIGURE 7-7. (a) Numerical integration, method I. (b) Numerical integration, method II.

measurement of the rectangles in the manner of the previous figure and add to it the area of the remaining triangle.

We must sometimes create special methods of integration, depending upon the situation. Consider the method called *geometric integration* (Figure 7-8) which is used in some *cardiac output computers*. A CO computer is used in hospitals to tell the doctor how much blood your heart is pumping per unit of time. The physician threads a multi-lumen catheter that has a thermistor tip through the veins of your arm and into the right side of the heart. The thermistor tip of this special catheter is placed on the "output" side of the heart, resting in the pulmonary artery. One of the lumens has an output port in the vena cava, outside the input side of the heart. If an injectate, such as iced or room temperature saline, is passed through the lumen into the blood stream, we can measure the cardiac output by measuring the integral of the temperature change at the output side (see *Introduction to Biomedical Equipment Technology* by J.J. Carr and J.M. Brown, John Wiley & Sons). But there is a significant

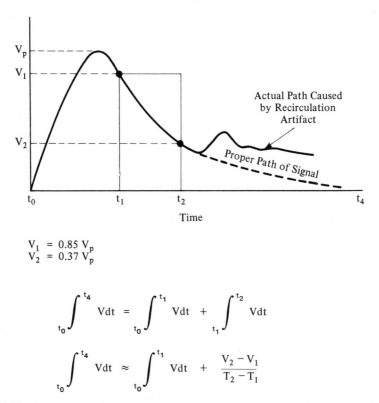

$V_1 = 0.85\ V_p$
$V_2 = 0.37\ V_p$

$$\int_{t_0}^{t_4} V dt = \int_{t_0}^{t_1} V dt + \int_{t_1}^{t_2} V dt$$

$$\int_{t_0}^{t_4} V dt \approx \int_{t_0}^{t_1} V dt + \frac{V_2 - V_1}{T_2 - T_1}$$

FIGURE 7-8. Geometric integration of exponential curves.

Section 7.2 / INTEGRATION AND DIFFERENTIATION

problem in some instruments: the *recirculation artifact*. Some of the cooled blood on the output side of the heart will recirculate and be "counted" again (see Figure 7-8). We can use geometric integration to overcome this problem—provided that the curve is exponential.

Figure 7-8 shows a typical temperature curve for a cardiac output computer. The period from t_0 to t_1 is a non-exponential section, so it is integrated in the normal manner [Figure 7-7(a) and 7-7(b)]. But the latter half of the waveform is exponential decaying. We can assume from empirical observations that the waveform enters the exponential section when the voltage decays to 85 percent of the peak value. Time t_1, therefore, is the onset of the exponential portion of the waveform. We can construct a rectangle that has an area that *approximates* the area of interest under the exponential section of the waveform. We take two samples of the voltage, V1 and V2. Sample V1 is taken at the onset of the exponential waveform, while V2 is taken when the voltage decays to the one time constant value, i.e., 37 percent of the V1 value. The total area is obtained by adding the area of the resultant rectangle to the area of the portion from t_0 to t.

8
Address Decoders

8.1 OBJECTIVES

1. To design simple eight-bit address decoder circuits.
2. To design *bank select* decoders.
3. To design I/O port decoders.

8.2 SELF-EVALUATION QUESTIONS

Before studying the material in this chapter, try answering the questions given below. These questions test your knowledge of the subject matter. If you cannot answer any particular question, then place a check mark beside it and look for the answer as you read the text.

1. How many bits of the address bus are required to decode the memory address H80 00?
2. Design a simple circuit to decode memory location H80 00 by a Z-80.
3. Do number 2, using a 6502 microprocessor instead of the Z-80.
4. How many *inverters* are required to decode I/O port address H62 using a 7430 NAND gate?

8.3 ADDRESSING IN MICROCOMPUTERS—A REVIEW

Microcomputers must be able to identify uniquely *locations* in which to *store data*. Some of these locations are memory locations, while others are I/O ports. In the 6502 uP, the I/O ports are treated as memory locations. Since the 6502 device has a 16-bit address bus, it can uniquely address up to 2^{16}, or 65,536 different locations. Allocations of specific addresses to either memory of I/O functions is determined by the designer of the 6502-based computer.

The Z80 uP device has a 16-bit address bus, so it too can address up to 65,536 different memory locations. In both the 6502 and Z-80 devices, each memory location will store a single eight-bit, or one-byte, word.

The Z80 differs from the 6502 in that it also provides eight-bit I/O port addressing capability. During the execution of an I/O instruction, the address of the I/O port appears on the low order byte of the two-byte address bus. The contents of the accumulator register in the CPU are passed over the high order byte of the address bus during this operation. Eight-bit I/O port coding means 2^8, or 256, unique I/O ports can be designated.

8.4 EIGHT-BIT DECODERS

An eight-bit address can uniquely designate either 256 memory locations or the same number of I/O ports. Many microcomputers use eight-bit address decoders for the following purposes:

1. To uniquely decode memory locations H00 to HFF.
2. To uniquely decode I/O ports H00 to HFF.
3. With a second eight-bit decoder, to uniquely decode memory locations H00 00 to HFF FF (a full 64 kbytes).
4. With *bank select* decoders, to select higher order banks of 256 memory locations each.

In order to design an address decoder, it is necessary to identify IC logic devices that will respond appropriately to the logic levels presented on the address bus. In most practical applications, the devices selected will include NAND gates, NOR gates, binary word comparators (e.g., 7485) and inverters. The decoder must be designed according to the memory or I/O devices that are being addressed. Most memory devices (but not all) have active-LOW *chip enable* (CE) terminals. This designation means that the IC turns on, i.e., becomes active, when the CE line is lOW, and is inactive when the CE line is HIGH. A 74100 TTL device used as an I/O port register, on the other hand, is an active-HIGH device.

In practical microcomputers with 1024 or fewer memory locations, the memory chip(s) will have their own internal address decoders, so they will be directly addressed from the address bus. No external address decoders are needed in that case (i.e., item 1, above). The remaining three applications, however, will require the generation of a decoded *select* signal.

The first address decoder is shown in Figure 8-1, and is one of the most popular circuits used. The 7430 IC is a TTL eight-input NAND gate. Its output will remain HIGH as long as any one of the eight inputs is LOW. The output terminal of the 7430 will go LOW only if all eight inputs are HIGH. The trick is to make the entire set of inputs HIGH when the correct address is present on the lower eight bits of the address bus. Of course, if the address is FF (hex) (11111111 in binary), then we have no problem. Connecting one each of the 7430 inputs to one of the

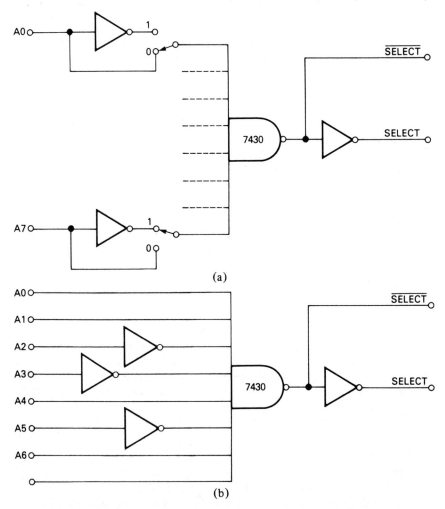

FIGURE 8-1. (a) Switches select whether "1" or "0" is required to generate a SELECT pulse. Only the LSB and MSB positions are shown in detail; (b) decoder for address selected.

Section 8.4 / EIGHT-BIT DECODERS

lower-order bits of the address bus will automatically give us our decoder. But all other addresses will require one or more inverters between the address bus lines and the inputs of the 7430. If you want maximum flexibility, then one inverter may be dedicated to each 7430 input. But this is a terrible waste of inverters, because there is only one address in which all of the inverters are required: 00000000 (binary)! All other addresses will use fewer than eight inverters. As a practical matter, most commercially available I/O printed circuit boards have but three or four inverters. The user is then asked to select I/O port addresses carefully so that no more than three or four zeroes occur. It is rare indeed that all 256 possible I/O ports would be required, so this is not the sacrifice that it might appear. In the example of Figure 8-1 we have shown inverters only on the A0 and A7 lines, with those for the other lines implied. In actual practice, most designers have the inverters wired in with small jumpers, rather than formal switches, so that they may be dedicated to *any* 7430 input required.

By way of illustration, let us assign the address 11010011 (D3 in hex) to an I/O port. We see by inspection that all but three of the bits in this address are ones, so no inverters will be needed for them. Only the zero bits (i.e., A2, A3, and A5) will require inverters. The decoder for this address is shown in Figure 8-1(b). Notice that the A0, A1, A4, A6, and A7 lines are connected directly to 7430 inputs, while the A1, A2, and A5 address lines are passed through inverters before being applied to the 7430 inputs. When the address 1 1 0 1 0 0 1 1 appears on the bus, then, all of the 7430 inputs will see ones, and the 7430 can drop LOW. This creates a SELECT signal for use by the I/O circuitry. An optional inverter will turn this signal upside down, creating a SELECT signal, for those cases where a positive-going transition is needed.

Another form of address decoder is shown in Figure 8-2. This circuit

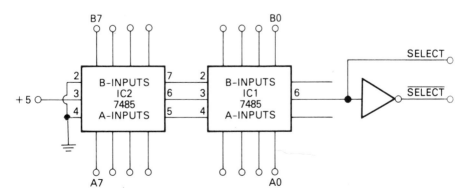

FIGURE 8-2. Using 7485 TTL comparators as an address decoder.

is based on the TTL-type 7485 comparator IC. This device will compare two four-bit words ("A" and "B") and issue an output that indicates whether A is equal to, greater than, or less than, B. Of these, we are interested in the A = B output (pin 6). The 7485 has cascading inputs that sense the status of a lower-order four bits. We need two 7485 devices connected in this cascade manner in order to use it to decode an eight-bit address bus.

We connect the bits of the address bus to the A-inputs of the 7485's. The B-inputs are used to program the device with the address of the port to be selected. In the previous case, we selected port D3 (hex), i.e., 11010011 in binary and 211 in decimal, using a 7430. If we wanted to use the 7485's as shown in Figure 8-2, we would program IC1 with the binary word 0011 (i.e., 3 hex), and IC2 with 1101 (i.e., D hex). When this address appears on the address bus, the A = B output of $\overline{\text{IC1 will}}$ go HIGH, forming a SELECT signal. An inverter is needed if a $\overline{\text{SELECT}}$ is desired instead.

We can also use any of the various "1-of-N" decoder ICs as address selectors. the 7442 is a 1-of-10 decoder, while the 74154 device is a 1-of-16 device. Each of these has a four-bit binary input to determine which output line (0 through 9, or 0 through 15) will go LOW. Figure 8-3 shows the use of the 7442 1-of-10 decoder. Two 7442 devices are needed. A NOR gate is connected so that one input of the NOR gate is driven by one of the outputs of each 7442.

As an example, let us say that we want to use the circuit of Figure 8-3 to decode address 115 (decimal), which is 73 in hex. The binary code for "7" is 0111, and the code for "3" is 0011. We want, then to see binary code 01110011 on the A0–A7 lines. We connect the four-bit inputs of IC1 to the lower-order four bytes of the address line (A0–A3), and the high-order four-bits (A4–A7) to the four-bit inputs of IC2. One input of the NOR gate is then connected to the "3" output of IC1, and the other input is connected to the "7" output of IC2. When the correct

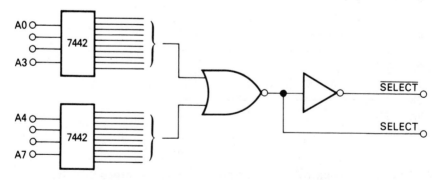

FIGURE 8-3. Using 7442 1-of-10 decoder as an address decoder.

Section 8.4 / EIGHT-BIT DECODERS

address appears, both of these outputs will drop LOW, causing the output of the NOR gate to snap HIGH. This signal then becomes our SELECT signal. Again, an inverter is used to form a $\overline{\text{SELECT}}$ signal where required.

Address Block Decoding

Most microcomputers use more than 1K of memory, yet many of the memory chips available are only 1024-byte (with some being 256-byte). Although there are more modern devices capable of very large byte arrays, many users still prefer the older, smaller devices. The question arises, "How does the memory device allocated to a location greater than the maximum address in each individual chip know when it is being addressed?" The solution seems to be ordering of the memory in 1K blocks, and then the use of some form of address decoding to tell which 1K block is being designated.

Figure 8-4 shows a selection scheme used by several manufacturers of 8K memory banks. Each block of this memory is an array of 1024 bytes, so every location can be addressed by bits A0–A9 of the address bus. The address pins for all devices are connected together to form the address bus (A0–A9). We must, however, select which of the eight blocks is addressed at any given time. One way to do this is to use a data selector IC. The 7442 device shown in Figure 8-4 is a BCD-to-1-of-10 decoder. It will examine a four-bit binary (i.e., BCD) input word, and issue an output condition that indicates the value of that word. In this simplified example, we are going to limit the memory size to 8K, so only the 1, 2, and 4 inputs of the 7442 are needed. The input weighted 8 is grounded (i.e., set = 0). The 7442 indicates the active output by going LOW, exactly the right condition for the RAM devices in the memory blocks. The table below shows the code that will exist on the A10–A12 bits of the address bus for the various memory addresses in the range 0–8K.

MEMORY LOCS.	A13	A12	A11	A10	BLOCK NO.	7442 OUTPUT	7442 PIN
0K-1K	0	0	0	0	0	0	1
1K-2K	0	0	0	1	1	1	2
2K-3K	0	0	1	0	2	2	3
3K-4K	0	0	1	1	3	3	4
4K-5K	0	1	0	0	4	4	5
5K-6K	0	1	0	1	5	5	6
6K-7K	0	1	1	0	6	6	7
7K-8K	0	1	1	1	7	7	9

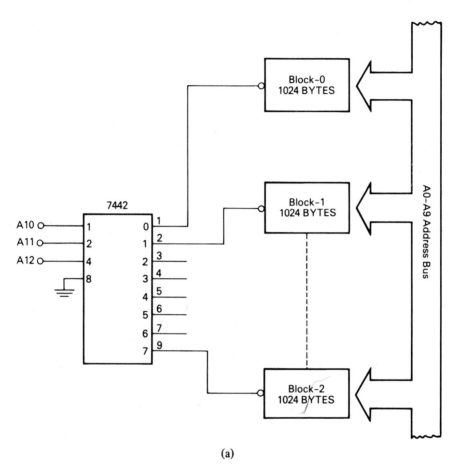

FIGURE 8-4. (a) Using 7442 in bank selection of memory; (b) code for above.

Section 8.4 / EIGHT-BIT DECODERS

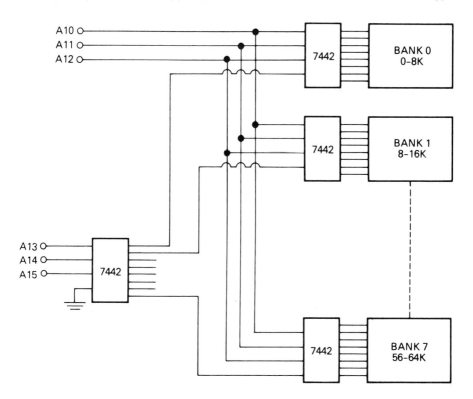

FIGURE 8-5. Multiple bank selection.

For an 8K memory, then, the lower 10 bits of the address bus (A0–A9) select which location in the individual chips is wanted, and A10–A12 select which block of 1024 bytes contains the address.

In the example of Figure 8-4 we limited the memory size to 8K. This was done intentionally to keep the circuit simple. But how do we select memory in ranges higher than 8K? The answer is to use the 7442 input weighted "8" as a bank select control. Recall from Figure 8-4 that this input was kept grounded. If it is HIGH, then none of the eight outputs of the 7442 will go LOW. But if it is LOW, then the circuit will work. Figure 8-5 shows a simplified selection scheme for all 65K, using the "8" weighted inputs of the 7442 block selectors as a *bank select* terminal. Each bank of 8K contains its own block select 7442, and one additional 7442 is used to select the bank of 8K that will become active. The table below shows the codes existing on address lines A13–A15 for each 8K bank of locations:

MEMORY LOCS.	BANK NO.	A15	A14	A13	7442 OUTPUT	7442 PIN LOW
0K-8K	0	0	0	0	0	1
8K-16K	1	0	0	1	1	2
16K-24K	2	0	1	0	2	3
24K-32K	3	0	1	1	3	4
32K-40K	4	1	0	0	4	5
40K-48K	5	1	0	1	5	6
48K-56K	6	1	1	0	6	7
56K-64K	7	1	1	1	7	9

Figure 8-6 shows an alternate bank selection circuit that is based on a three-input NAND gate (i.e., one section of a 7410 TTL IC device). The properties of a NAND gate are

1. If any input is LOW, the output is HIGH.
2. If *all* inputs are HIGH, then the output is LOW.

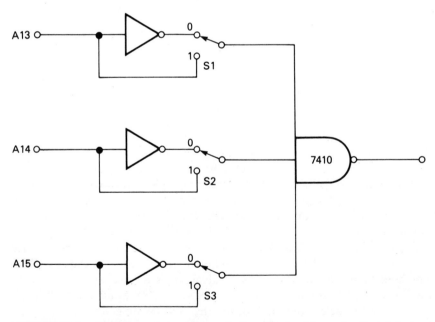

FIGURE 8-6. Switches determine whether logical-1 or logical-0 is required.

Section 8.4 / EIGHT-BIT DECODERS

In this case, then, all three of the inputs must be HIGH for the output to drop LOW. If the output of the NAND gate is used to drive the "8" input of the 7442, then the particular bank served by that 7442 will be selected only when all three inputs are HIGH.

How do we contrive the circuit to force all inputs HIGH only when the correct bit pattern is seen on lines A13-A15? The solution is the inverters and switches shown in Figure 8-6. Each switch selects either the inverted (i.e., "0" position) or noninverted (i.e., "1" position) versions of each address bus signal. We set the three switches according to our bank selection format, using the codes from the table given previously. Each switch is to be set to the position corresponding to the digit expected at that input when the address bus code is correct. For bank 0, for example, the code is 000. If S1-S3 are set to "0" position, then the NAND gate sees the inverted address line signals. When 0 0 0 appears on A13-A15, the NAND gate sees 111. Since this is the condition required, the output drops LOW and turns on the selected bank.

Note that Intel manufactures a 1-of-8 decoder intended specifically for bank selection in the 8080A device. It should also work nicely with the Z80 and other devices.

9
Interfacing Memory

The typical microprocessor chip uses a 16-bit address bus, so it is able to directly address up to 2^{16}, or 65,536 memory locations. Note that this upper limit is usually written "64K" rather than "65K" because a "computer-K" is 1024 instead of 1000. The data bus uses one byte (eight bits), so each memory location can store a single eight-bit word.

The mixture of possible memory devices used with the uP includes static random access memory (RAM), dynamic RAM, read only memory (ROM), programmable read only memory (PROM), erasable PROM (EPROM), plus a number of devices such as analog-to-digital converters (ADC), and digital-to-analog converters (DAC), which are sometimes treated as memory. This technique, called *memory mapping,* makes some data acquisition chores easier (or at least faster).

9.1 CONTROL SIGNALS FOR MEMORY OPERATIONS

Let us consider the Z80 as our interfacing example. We must be cognizant of the basic control signals that apply to memory operations: $\overline{\text{MREQ}}$, $\overline{\text{WR}}$, and $\overline{\text{RD}}$. These signals are the memory request, write, and read, respectively. The memory request signal will drop LOW whenever the CPU is executing either a memory read or memory write operation. It tells the system that the data on the bus are memory data. If a memory write operation is taking place, then the *write* ($\overline{\text{WR}}$) signal will also go LOW. If, on the other hand, it is a memory read, then the *read* ($\overline{\text{RD}}$) signal will go LOW. All memory operations, therefore, will generate a LOW on two control pins of the Z80: $\overline{\text{MREQ}}/\overline{\text{WR}}$ for memory write operations and $\overline{\text{MREQ}}/\overline{\text{RD}}$ for memory read operations.

Most integrated circuit memory devices have at least one chip enable (CE) pin, and some have two chip enable pins (labelled CE1 and

Section 9.1 / CONTROL SIGNALS FOR MEMORY OPERATIONS

CE2). There also may be a *read/write* (W/R) pin to instruct the device whether the desired operation is a memory read or a memory write.

One of the simplest cases is shown in Figure 9-1. Here we see 1024 bytes of *read only memory* (ROM) interfaced directly to the Z80. In this case, we have assigned the ROM to the lower 1K of the memory address range. The locations available, then, are 00 00 (H) to 03 FF (H). Since we are dealing with the lower 1K, we need only the lower-order byte of the address bus, A0-A7, plus the two least significant bits of the upper-order byte (A8 and A9).

Two chip enable (CE) terminals are available. We use one of them (CE2) to make sure that the ROM will respond only to addresses in the lower 1K of memory. Address bus bit A10 will always remain LOW when the CPU is addressing a location in the lower 1K, but will go HIGH when an address greater than 03 FF (H) is selected. The ROM, therefore, is enabled only when the address on the address bus is less than 03 FF (H).

The second chip enable pin (CE1) is used to turn on the ROM only when the memory read operation is taking place. This CE pin wants to see a HIGH for turn-on of the ROM. Recall that a NOR gate will output a HIGH only when both inputs are LOW. We can, therefore, create a device select command for CE1 by applying the $\overline{\text{MREQ}}$ and $\overline{\text{RD}}$ control signals from the CPU to the inputs of a NOR gate. CE1 will go HIGH, then, only when a memory read operation takes place.

At least two of the more popular ROM chips require only a single chip enable command. In the example shown in Figure 9-2(a), the chip enable is an active-LOW input (so is designated $\overline{\text{CE}}$). This terminal is brought LOW whenever we want to read the contents of one of the locations in the chip.

The example shown in Figure 9-2(a) is a 256-byte ROM, with a single $\overline{\text{CE}}$ terminal. We must, therefore, construct external circuitry that will bring the chip enable terminal LOW when we want to perform the read

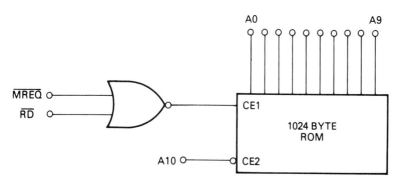

FIGURE 9-1. Operating CE of memory from $\overline{\text{MREQ}}$ and $\overline{\text{RD}}$.

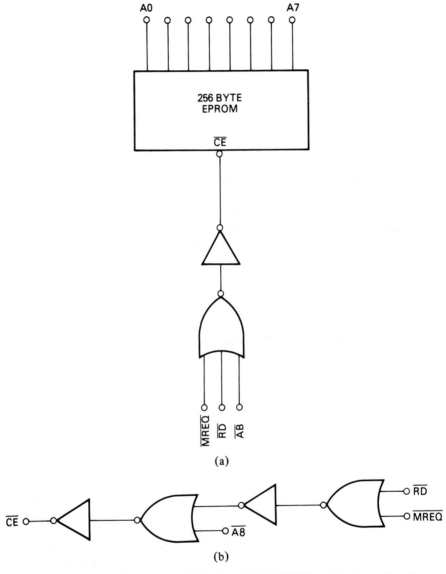

FIGURE 9-2. (a) Enabling EPROM from MREQ/RD/A8; (b) Same function accomplished with two-input gates.

operation. The simplest way is to use a three-input NOR gate and an inverter. The output of the NOR gate will go HIGH only when all three of the inputs are LOW. We connect the $\overline{\text{MREQ}}$, $\overline{\text{RD}}$, and bit A8 of the address bus to the respective inputs of the NOR gate. When the conditions are met, then the output of the gate snaps HIGH and is then inverted to become the $\overline{CE}$ signal required by the EPROM chip.

As an alternative method is shown in Figure 9-2(b). Here we are using two inverters and a pair of NOR gates to form the $\overline{CE}$ signal. The idea is to cause $\overline{CE}$ to go LOW when the three conditions are met. To do this, we must see both inputs of NOR gate G2 LOW simultaneously. One of the inputs is connected to bit A8 of the address bus, while the other is connected to the inverted output of NOR gate G1. The inputs of G1 are, in turn, connected to the $\overline{MREQ}$ and $\overline{RD}$ signals.

A situation that is a little more complicated is shown in Figure 9-3. Here we are interfacing static RAM devices that have a chip enable and a $\overline{R/W}$ terminal. This latter terminal will cause the device to read out data when LOW and allow writing in data when HIGH. We connect the $\overline{R/W}$ terminal, then, to the $\overline{RD}$ signal of the Z80 CPU.

The chip enable in this example wants to see a HIGH in order to turn on the device. We can, then, correct CE to the output of a NOR gate. The $\overline{MREQ}$ and $\overline{A8}$ signals are connected to the two inputs of the NOR gate. If both of these signals go LOW simultaneously, and the $\overline{RD}$ is also LOW, a memory read operation takes place from the location addressed by A0-A7. Alternatively, if the $\overline{MREQ}$ and $\overline{A7}$ signals are LOW, and the $\overline{RD}$ signal is HIGH, then a memory write operation will take place.

Note in Figure 9-3 that two chips are used to form a 256-byte static RAM memory. Most memories require more than a single chip in order to form a complete byte-array. In this case, each memory chip contains a 256 × 4-bit array, so two connected together will form a 256 × 8-bit array (i.e., 256 bytes of memory). The popular 2102 device is listed as a 1024 × 1-bit device. Connecting eight of these devices into an array will result in a 1024-byte memory.

9.2 DYNAMIC MEMORY

Dynamic memory (RAM) will not hold its data for an indefinite length of time unless a refresh operation is performed. The refresh operation is a function of the CPU in most cases, although some non-CPU example exist. Although the use of static RAM will eliminate this problem, it will do so only at the cost of a higher power consumption. The Z80A device provides for refresh of the dynamic memory by adding a refresh segment to the M1 (instruction fetch) machine cycle.

During clock periods T3 and T4 of the M1 cycle, used by the Z-80 for the decoding of the instructions fetched in the earlier T-periods, a refresh signal is generated. The $\overline{RFSH}$ terminal (pin 28) of the Z80 will go LOW during this period. Note that this signal must be used in conjunction with

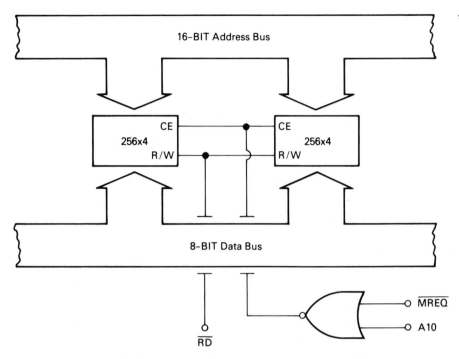

FIGURE 9-3. Interfacing RAM.

the $\overline{\text{MREQ}}$ (memory request) signal, because the $\overline{\text{RFSH}}$ is guaranteed to be stable only when the $\overline{\text{MREQ}}$ is also active.

During the refresh period the lower portion of the address of a refresh location is placed on the lower seven bits (A0-A6) of the address bus (A7 is 0). The data on A0-A6 are from the R register in the Z80, which is incremented after each instruction fetch. The upper eight bits of the address bus carry the contents of the I register. Figure 9-4 shows an example of an 8K dynamic RAM interfaced to a Z80. In this particular case, 4K × 8-bit dynamic RAMs are used. If no other RAM is used, we may use bit A12 of the address bus as a chip-select line.

9.3 ADDING WAIT STATES

All solid-state memory chips require a certain minimum period of time to write data into, or read data from, any given location. Many such devices are graded (and priced) according the memory speed. The popular 2102 device, a 1K × 1-bit IC, is available in 250-nanosecond, 400-nanosecond,

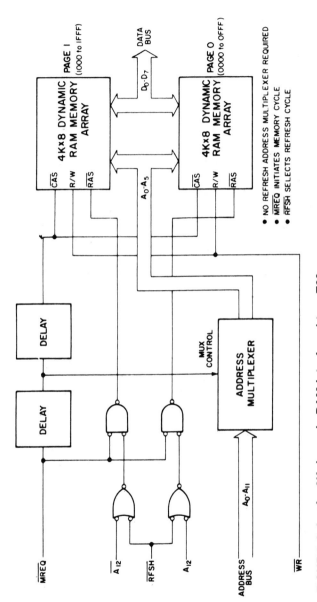

FIGURE 9-4. An 8K dynamic RAM interfaced to a Z80.

and 500-nanosecond versions. Of course, the cost per chip rises with the speed.

Since the Z80A can operate at speeds up to 4 mHz, we sometimes find the cycle (M1 or memory) over before the data have settled to, or from, memory. This problem can be overcome by adding the circuitry shown in Figure 9-5. Both of these circuits generate a $\overline{\text{WAIT}}$ input (pin 24 of the Z80) equal to the period of one clock pulse.

The circuit in Figure 9-5 uses both sections of a TTL 7474 dual Type-D flip-flop. The 7474 is a positive-edge triggered device, meaning that data on the D-input are transferred to the Q output *only* during the positive-going transitions of the clock pulse.

Immediately after the onset of clock pulse T1, the $\overline{\text{M1}}$ line goes LOW, forcing the D input of FF1 LOW. When clock pulse T2 snaps HIGH, then, this LOW is transferred to the Q output of FF1. This signal becomes the $\overline{\text{WAIT}}$ signal for the CPU, and inserts one additional clock period (T_w) into the M1 cycle.

At the onset of clock period T_w, then, FF2 sees a LOW (i.e., the $\overline{\text{WAIT}}$ signal) on its D input. This LOW is transferred to the Q output of FF2. The Q_2 terminal (FF2) is connected to the *set* input of FF1, so this condition forces the Q_1 (FF1) HIGH again, thereby terminating the action.

A similar circuit, shown in Figure 9-5, is used to add a wait state to any memory cycle. When the first clock pulse (T1) arrives, the $\overline{\text{MREQ}}$ line goes LOW, forcing the D-input of FF1 LOW. At the onset of clock pulse T2, then, this LOW is transferred to the Q output of FF1. At this time $\overline{Q_1}$ is HIGH and Q_2 is HIGH, so the output of the NAND gate drops LOW. (Both NAND inputs must be HIGH for the output to be LOW.) This causes the $\overline{\text{WAIT}}$ input of the CPU to become active. But at the onset of T_w, the added clock period, the LOW on Q_1 is transferred to Q_2. This forces one input of the NAND gate HIGH, thereby cancelling the $\overline{\text{WAIT}}$ signal.

9.4 MEMORY-MAPPED DEVICES

Some peripheral devices used with microcomputers can be more efficiently employed if they are treated as a memory location, instead of an I/O device. An example might be a *digital-to-analog converter* (DAC), which is a device that creates an analog output voltage (or current) that is proportional to a binary digital word applied to its input.

Figure 9-6 shows how an eight-bit DAC can be interfaced with a Z80 *as if the DAC were a memory location*. The DAC requires stable input data, but the data on the bus are transitory. Therefore, we need a *data*

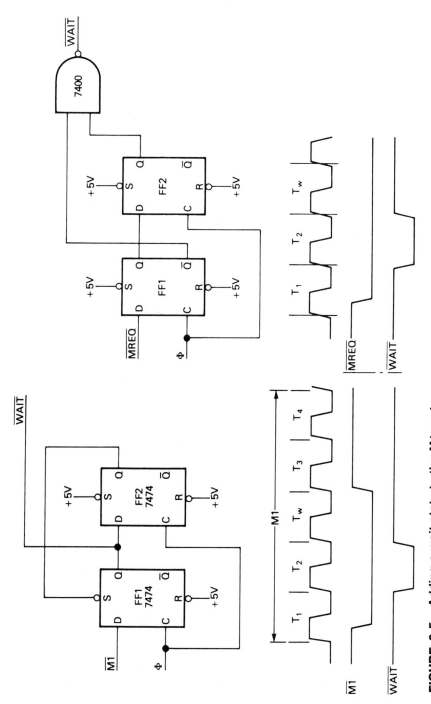

FIGURE 9-5. Adding a wait state to the M1 cycle.

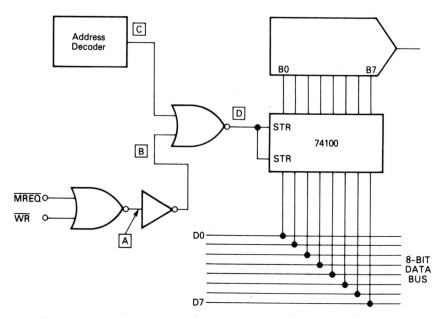

FIGURE 9-6. Using a device such as a DAC as a piece of memory.

latch between the 8-bit data bus and the DAC inputs. There are a number of interface chips that will perform this job, but most of those special-purpose devices are costly. A low-cost solution, which works just as well, is to use a 74100 TTL dual quad-latch. The two four-bit sections of the 74100 become an eight-bit latch when the *strobe* terminals are tied together.

The 74100 latch transfers the information on the data bus to the DAC when the strobe line is HIGH. The 74100 outputs, connected to the DAC inputs, will retain these data when the strobe line again goes LOW. The idea, then, is to make the 74100 strobe line HIGH during the period when the desired DAC input data are present on the data bus.

Three criteria must be met before the data on the bus can be input to the DAC: (1) The write signal ($\overline{WR}$) must be active; (2) the memory request ($\overline{MREQ}$) must be active; (3) the correct address (the address of the location assigned to the DAC) must be present on the address bus. The first two criteria are examined by a single NOR gate. When both $\overline{WR}$ and $\overline{MREQ}$ are LOW (i.e., active), we are producing a memory write operation. This will cause point "A" to go HIGH, and point "B" to go LOW. We do not want the DAC to respond, however, unless point "C" is LOW at the same time. When point "C" is LOW, we know that the address for the DAC is being sent over the address bus. When all three criteria are

Section 9.4 / MEMORY-MAPPED DEVICES

met, the strobe input of the 74100 (point "D") will go HIGH. This will allow transfer of data from the data bus into the DAC.

Most microcomputers have less than the full 64K complement of memory. This is why most memory-mapped devices tend to be allocated addresses in the upper 32K of memory. This, incidentally, allows us to use bit A15 of the address bus to discriminate between the various addresses.

10
Interfacing I/O Ports

10.1 OBJECTIVES

1. To design parallel input/output ports.
2. To design serial TTL, RS-232 and current-loop input/output ports.
3. To use peripheral interface devices to provide I/O capability.

10.2 SELF-EVALUATION QUESTIONS

Before studying the material in this chapter, try answering the questions given below. These questions test your knowledge of the subject. If you cannot answer a question, then place a check mark beside it and look for the answer as you read the text.

1. How can a 74100 device be used to make an eight-bit output port?
2. The Intel 8212 device can be used as a _____ data register for I/O applications.
3. Why are tri-state output devices required for microcomputer input registers?
4. What modes are available for the 6522 interface device?

10.3 I/O PORTS

Input and output ports (collectively, I/O ports) are used to transfer data to and from the world outside of the computer. An input port is used to transfer data *from* the outside world to the CPU, while the output port is used to transfer data *to* the outside world from the CPU.

Data transferred in an I/O port operation must typically go through

Section 10.4 / GENERATING PORT/DEVICE SELECT SIGNALS

the CPU accumulator register. This limitation differentiates I/O operations from those operations that transfer data directly from the CPU registers to memory locations or those that read or write data directly from the outside world to memory locations. This latter operation is known as *direct memory access,* or DMA.

Some microprocessor devices provide for separate I/O ports, while others use a *memory mapped* system for the I/O ports. In the latter type of system, the I/O ports are treated as memory locations by the CPU. An input operation is then treated as a memory read operation from the location assigned by the designer as a port. Of course, each memory location specified as an I/O port is one less location that is available for data.

The easiest I/O to design is the ordinary *parallel port,* in which all bits are transferred to or from the data bus *simultaneously.*

The *serial port* is designed to transfer the data *one bit at a time.* The serial I/O port is a little more difficult to design than parallel ports, but permits the most efficient transfer of data over a communications link. To transmit eight bits of data in parallel form would require eight separate telephone lines or radio channels, obviously too costly for most applications.

There are various ways to design serial I/O ports, and several standards that could possibly apply. In this chapter, we will consider the most common methods.

10.4 GENERATING PORT/DEVICE SELECT SIGNALS

Computers based on microprocessors that provide separate I/O function generally require a means to *decode* the *port address* and sense whether an *input* or *output* operation is to take place. The Z80 microprocessor, for example, uses the following signals to define I/O operations: (a) an eight-bit address on the lower byte (A0-A7) of the 16-bit address bus to specify the *port number,* (b) an active-LOW input/output request ($\overline{IORQ}$) line to indicate that an I/O operation is taking place, and (c) active-LOW *read* ($\overline{RD}$) and *write* ($\overline{WR}$) lines to indicate whether the request is an input or an output, respectively.

Figure 10-1 shows a method for generating an active-HIGH OUT or IN signal in Z80 based microcomputers. This circuit is based on a single 7402 IC (quad two-input NOR gate). Gates G3 and G4 are used to create the OUT and IN signals, and are controlled by the $\overline{IORQ}$ signal and the signals from *read* and *write* lines. Recall the rules governing NOR gate operation: a HIGH on any input produces a LOW output, and it requires all inputs be LOW for the output to be HIGH. When the $\overline{IORQ}$ line is inactive (i.e., HIGH), then the NOR gate outputs (G3 and G4) are forced

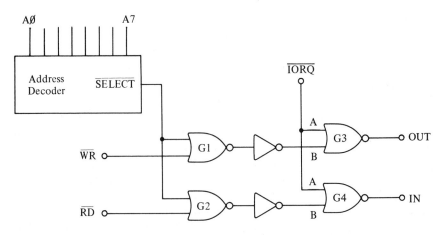

FIGURE 10-1. Simple IN and OUT pulse generator.

LOW. If, however, IORQ becomes active (i.e., LOW), then the NOR gate outputs are unlatched and will be HIGH or LOW depending upon the state of the RD and WR lines.

Case no. 1: Input operation. An input port operation is a read operation because data is transferred *into* the accumulator. When the Z80 device executes an input operation, therefore, the IORQ and RD go LOW. A LOW condition on IORQ unlocks gate G4 by bringing input-A LOW. The RD line, and the SELECT line from the address decoder, controls gate G2. When *both* of these lines are LOW, then, input-B of G4 will also go LOW. We now find that both inputs of G4 are LOW, so the IN line (G4 output) goes HIGH.

Case no. 2: Output operation. An output operation is a write function because data is transferred *from* the accumulator to the outside world. When a Z80 microprocessor executes an output operation, therefore, the IORQ and WR go LOW. A LOW condition on IORQ unlocks gate G3 by bringing input-A LOW. The WR line and the SELECT line from the address decoder controls gate G1. When both of these lines are LOW, then input-B of G3 will also go LOW. We now find both inputs of G3 LOW, so the OUT line goes HIGH.

A *select* circuit based on the 7442 TTL IC is shown in Figure 10-2. The 7442 is a TTL *BCD-to-1-of-10 decoder*. It will examine the four-bit binary coded decimal (BCD) input line and generates a unique active-LOW output that depends on the decimal equivalent of the input word.

Section 10.4 / GENERATING PORT/DEVICE SELECT SIGNALS

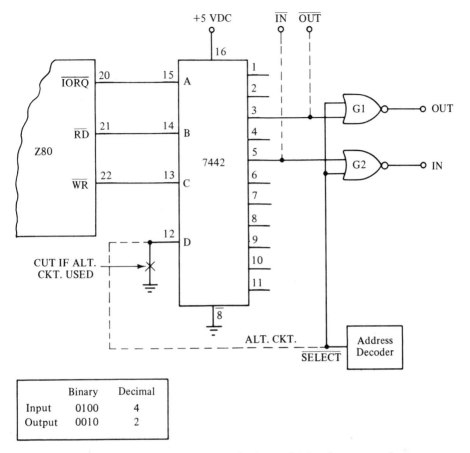

FIGURE 10-2. Using the 7442 as a device select pulse generator.

For example, when the binary word 0100 (4_{10}) appears on the inputs, then the "4" output (pin no. 5) goes LOW.

Two methods are shown in Figure 10-2 for using the 7442 in this application. The main circuit is shown in solid lines, while the alternate circuit is shown in dotted lines.

The main circuit in Figure 10-2 works by considering the $\overline{IORQ}$, $\overline{RD}$, and $\overline{WR}$ lines from the Z80 as a three-bit binary word, and these lines are connected to the *A, B,* and *C* lines of the 7442, respectively. The *D* input line of the 7442 is wired permanently LOW. An input operation causes the $\overline{IORQ}$ and $\overline{RD}$ to be LOW, so the word applied to the 7442 is 0100, or decimal 4. This situation causes pin no. 5, which is the "4" output, to go LOW. Similarly, an output operation causes $\overline{IORQ}$ and $\overline{WR}$ to be LOW, so the word applied to the 7442 is 0010, or decimal 2. This condition causes pin no. 3 to go LOW during the output operation.

Like the previous circuit, Figure 10-2 is used with specific I/O ports, so it must recognize the signal from the address decoder. The $\overline{\text{SELECT}}$ line will go LOW when the CPU selects its address; at all other times this line is HIGH. The $\overline{\text{SELECT}}$ line is connected to both G1 and G2.

When pin no. 3 of the 7442 goes LOW, and the $\overline{\text{SELECT}}$ line is also LOW, then the output of gate G1 will go HIGH, providing an OUTPUT signal. This signal is used to tell the output port circuitry to accept data on the bus.

An INPUT signal is created by gate G2 under similar circumstances. The $\overline{\text{SELECT}}$ line will go LOW if the address for that port is indicated. Gate G2 knows that an input operation is demanded because pin no. 5 of the 7442 goes LOW in response to the binary code 0100 at its inputs.

In the alternate version of this circuit we can get rid of the two NOR gates by using the $\overline{\text{SELECT}}$ line from the address decoder to activate the D input of the 7442. This input wants to see a LOW condition before the IN and OUT lines are activated (again, pins 5 and 3 are used). If another I/O port is being called up, the A-B-C inputs will still see the correct code from the Z80 (010 and 100), but the D input sees a "1." The codes applied to the 7442 then are 1010 and 1100—so they will not activate any 7442 output line. When this port is selected, however, the 7442 input codes are either 0010 or 0100, so the appropriate output is activated.

Figure 10-3 shows a circuit that uses just two NOR gates to create the IN and OUT signals. The gates selected for this circuit are three-input

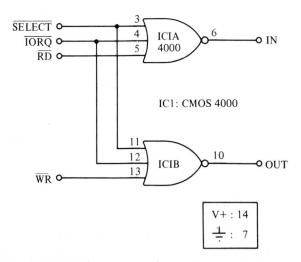

FIGURE 10-3. CMOS NOR gates used as a device select pulse generator.

NOR gate sections of a CMOS 4000 IC. This circuit is simple and will work with most microcomputer systems. It will not work, however, in high speed (10 mHz and up) clock systems. These NOR gates require all three inputs LOW for the output to be HIGH. The only time this situation will occur is during input operations for IC1A and output operations for IC1B.

Thus far, all of the circuits presented have been for specified I/O ports, i.e., an address decoder was included in the design. It is often desirable, however, to create *system I/O* signals that are active for any input or output operation. We then provide a simple address decoder at each port or on each printed circuit card within the microcomputer. This structure is popular in microcomputers that allow for future expansion through the addition of plug-in I/O cards. In a Z80 based system, we can create the system signals by adapting the circuit of Figure 10-3. We would either ground the 4000 device inputs used for the $\overline{\text{SELECT}}$ signal, or use two-input NOR gates such as the 7402 device.

10.5 PARALLEL OUTPUT PORT DESIGN

Although many microcomputers use special-purpose IC devices for I/O ports, it is often less costly to use ordinary IC devices from the regular digital logic families in these applications.

A Type-D flip-flop can be used to store one bit of data, so an array of eight Type-D flip-flops can be used to hold one byte of data. Figure 10-4(a) shows the symbol for the Type-D flip-flop. Data applied to the *D*-input will be transferred to the Q output only when the clock line is HIGH. If the clock line remains HIGH, then the data on the Q output will follow the data on the D input. If, however, the clock line is LOW, then the data on the Q output remains the same; i.e., the FF will hold the last valid D-input data present before the clock dropped LOW. The flip-flop will then *ignore* all further changes of the D-input data.

There are a number of Type-D flip-flops available in the major digital logic IC families. Certain devices called *data latches* contain four, or eight, Type-D flip-flops in arrays that share a common clock line. An example is the 74100 device shown as an output port register in Figure 10-4(b).

The 74100 is a dual four-bit data latch, and it contains a pair of Type-D flip-flop arrays—each containing four flip-flops. If the two clock lines (pins 12 and 23) are tied together, then the 74100 operates as an eight-bit register. The eight D-inputs are connected to the data bus, while the eight Q outputs are used as the output bits. This type of circuit forms a

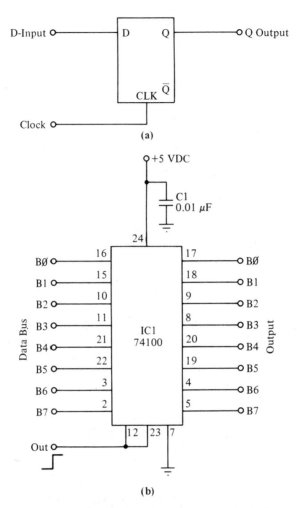

FIGURE 10-4. (a) Type D flip-flop. (b) Latched output port using 74100 dual quad-latch.

latched output because the data remains fixed on the output lines after the execution of an output instruction is completed.

The common clock line from the 74100 is connected to the OUT signal from the port select circuit (Sec. 10-4). The data on the data bus is transferred to the output side of the 74100 register whenever the OUT signal line goes HIGH.

10.6 PARALLEL INPUT PORT DESIGN

Ordinary IC devices may also be used in the design of input ports. There is, however, a design constraint that limits our selection of devices to those which provide *tri-state* output lines.

The microcomputer data bus serves several functions, including transfers to and from many memory locations and an I/O port. If any one device causes a permanent HIGH or LOW condition on any line of the data bus, then there will be massive errors in the data. All devices connected to the data bus must, therefore, be tri-state devices.

A tri-state device is one that becomes effectively disconnected from the output terminal when it is not active. All digital ICs are at least two-state, i.e., the output terminal sees either a low impedance to ground (LOW condition), or, a low impedance to +5 volts DC (HIGH condition). The tri-state device has a third state in addition to the regular states in which the output terminal sees a high impedance to both ground and +5 volts DC anytime the *chip enable* (CE or $\overline{CE}$) terminal is inactive. There are only a few devices from the regular TTL lines that meet this criterion, an example of which is the 74125 device.

The 74125 is a four-bit TTL device that contains four noninverting buffers each of which has its own active-LOW chip enable ($\overline{CE}$) line. When $\overline{CE}$ is LOW, the buffer acts like any TTL buffer, so the output data directly follows the input data. However, when the $\overline{CE}$ line is HIGH, the buffer output line goes to a high impedance, and the device is effectively disconnected from the output line.

Figure 10-5 shows a pair of 74125 devices used to form an eight-bit input port. Since an input port should not provide latching, we cannot use the 74100 device. All four $\overline{CE}$ lines on both 74125 devices are tied together to form a master $\overline{CE}$ line that turns the port on and off in response to the $\overline{IN}$ signal.

The $\overline{IN}$ line will normally be held HIGH until an input operation takes place from that port. At that time, the $\overline{IN}$ line drops LOW, turning on the 74125s. This action connects the bits of the input ports to the data bus as long as $\overline{IN}$ is LOW, so input data appears briefly on the data bus.

10.7 BIDIRECTIONAL PARALLEL I/O PORT DESIGN

The job of designing and interfacing I/O ports is made easier by the use of certain special purpose IC I/O devices. Some of these are general enough to be used with a wide variety of microprocessor chips. The 8216/8226 and 8212 devices are examples. Produced by Intel for use with the 8080a

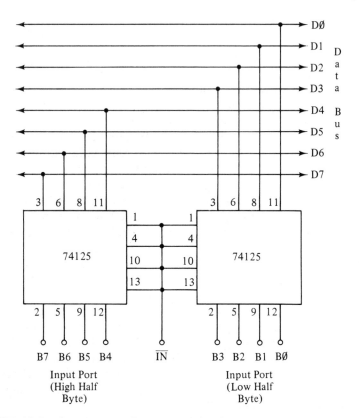

FIGURE 10-5. Input port using two 74125 chips.

microprocessor chip, they are frequently adapted for use with other CPU chips.

The 8216 and 8226 devices are almost identical to each other, except that the 8216 contains *noninverting* line drivers, while the 8226 contains *inverting* line drivers. Both are four-bit devices, so two are needed to drive an eight-bit data bus.

Figure 10-6 shows the logic diagram for the 8216 device. This same diagram is also used for the 8226, except that the line driver buffers are inverters. Lines DB0 through DB3 are connected to four bits of the data bus. The four input port lines are labelled DI0 through DI3, while four output port lines are designated DO0 through DO3.

The $\overline{\text{DIEN}}$ line controls the data direction. If this line is LOW, then the input lines are connected to the data bus. Alternatively, if $\overline{\text{DIEN}}$ is HIGH, then the output port is connected to the data bus.

The lines connected to the data bus have tri-state outputs, so they can be made to float at high impedance when not needed. The chip select

Section 10.7 / BIDIRECTIONAL PARALLEL I/O PORT DESIGN

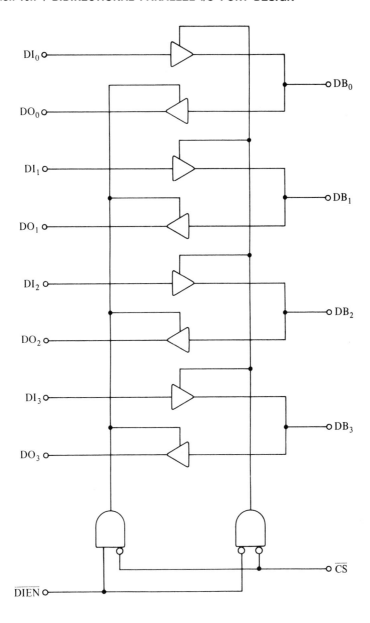

FIGURE 10-6. 8216 I/O port.

$\overline{\text{(CS)}}$ is an active-LOW terminal that controls the tri-state outputs. If $\overline{\text{CS}}$ is HIGH, then the outputs have a high impedance and will not load the bus. However, when $\overline{\text{CS}}$ is LOW, the output lines become active.

Only the $\overline{\text{DIEN}}$ line of the 8216/8226 devices affects the input function. To make an output port active, though, requires a HIGH on $\overline{\text{DIEN}}$ *and* a LOW on $\overline{\text{CS}}$.

The actual circuit configuration will depend upon the use of the chip (I or 0) and the microprocessor used. In a Z80 system, for example, we could connect $\overline{\text{CS}}$ to $\overline{\text{IORQ}}$ and $\overline{\text{DIEN}}$ to $\overline{\text{RD}}$ (which will be LOW for input operations and HIGH for output operations).

The Intel 8212 device is shown in Figure 10-7(a). This chip is an eight-bit directional I/O port that was designed to be compatible with the 8080a microprocessor. It also has been used with many other microprocessor chips. The 8212 also differs from the previous example in that it is (a) unidirectional (two are needed for a complete I/O), and (b) it contains latches so it will retain the last valid data. The latter feature is necessary for most microcomputer applications. An example showing a pair of 8212 devices in a bidirectional I/O port circuit is given in Figure 10-7(b).

10.8 SERIAL I/O PORTS

Parallel I/O ports prove to be expensive when data must be communicated over a long distance to another device. Unless the communications link is only a few feet in length, serial communications is cheaper. Transmission of eight data bits in parallel, for example, requires at least eight lines, and usually a ninth line is needed as a common or ground. The excessive cost of radio and telephone links should be obvious. In contrast, a serial communications link requires only one channel, be it local wire, telephone lines, or a radio channel.

In addition, there are numerous peripherals, such as teletypewriters, that use serial inputs or provide serial outputs. Serial data ports are slower than parallel ports because the data must be transmitted one bit at a time. The bit length is also greater because certain *format bits* must be added, i.e., start bit plus one or more stop bits.

Software Serial I/O Ports

Some designers use software to force a single bit of a parallel I/O port to function as if it were a serial port. In one simple method, the output word is loaded into the accumulator, and then the *shift* instructions are used to output the word bit by bit.

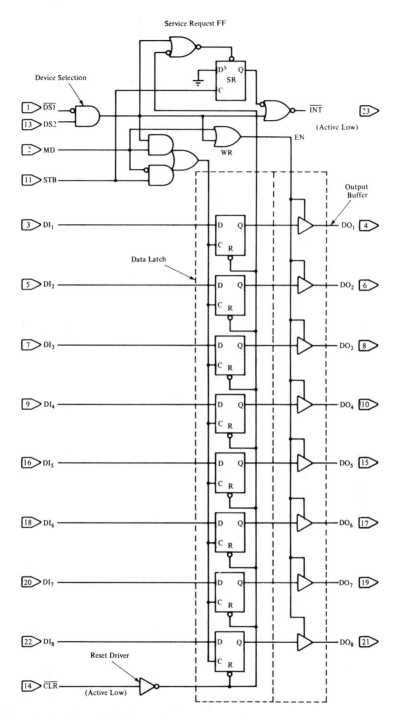

FIGURE 10-7(a). 8212 I/O port device.

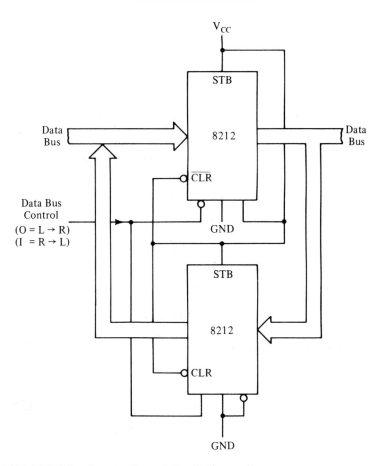

FIGURE 10-7(b). Connection of the 8212.

Another, more formal, technique is to write a software program to perform the logic functions of a *universal asynchronous receiver/transmitter,* or UART. Such programs are often designated *teletypewriter* or *printer driver routines.*

TTL/CMOS Hardware Methods

The basic circuit needed to make a serial output port is the *parallel-in serial-out* (PISO) shift register. Data from the accumulator is written to the parallel inputs of the shift register as if the register were either a memory location or a parallel output port. A program is then executed that will clock the flip-flop left to right, causing the shift register to output the data in serial format. Various clocking schemes are used, including successively generating an OUT signal, or writing to a specific (unused) memory location. In both cases, a device or location select pulse must be provided.

A serial input port can be built in a similar manner, but using a *serial-in parallel-out* (SIPO) shift register. Remember, tri-state outputs must be available to prevent loading of the data bus when the port is inactive.

The UART

Until recently the design of serial I/O ports was difficult and was considered at best a nettlesome chore for the digital design engineer. But today we have a specialized class of LSI integrated circuits that will do the job: the universal asynchronous receiver/transmitter, or UART.

Synchronous transmission requires that the clock pulses at both ends of the circuit be matched because data is transmitted at precise times. In general, the respective baud rates of the receiver and transmitter must be within ±0.01 percent of each other. Such precision adds complexity and cost to the system. We can reduce both complexity and cost, however, by using *asynchronous* transmission.

Figure 10-8 shows the block diagram for the now-standard TR1602A/B device. This UART is essential the same as the popular AY-1013 UART device. It is capable of either full duplex or half duplex operation because the receiver and transmitter sections are totally independent of each other, except for power supply connections.

The UART chip is particularly useful because it can be programmed externally for several different *bit lengths, baud rates, parity* (odd-even, receiver verification/transmitter generation), *parity inhibit,* and *stop bit* length (1, 1.5, or 2 stop bits). The UART also provides six different status flags: *transmission completed, buffer register transfer completed, received data available, parity error, framing error,* and *overrun error.*

The clock speed is 320 kHz (maximum) for the A and B versions, 480 kHz for the AO3/BO3 versions, 640 kHz for the AO4/BO4 versions, and to 800 kHz for the AO5/BO5 series. The receiver output lines are tri-state logic, so will float at a high impedance to both ground and the +5 volt line when inactive. The use of tri-state output allows the device to be connected directly to the data bus of a computer or other digital instrument.

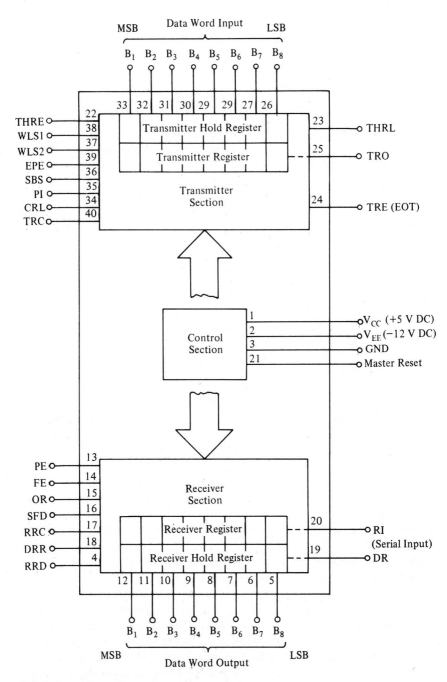

FIGURE 10-8. Block diagram of a UART.

Section 10.8 / SERIAL I/O PORTS

The transmitter section uses an eight-bit parallel input register that will accept data to be sent serially. It will convert the eight-bit data word received in the input register to serial format that includes the eight-bit word (also formatable to 5, 6, or 7 bits), start bit, parity bit and stop bits.

The receiver can be viewed as simply the mirror image of the transmitter. It receives a serial input word containing start bits, data, parity, and stop bits. This serial stream of data is checked for validity by comparison with parity and for the existence of the stop bits.

The UART data format (serial) is shown in Figure 10-9. The transmitter output pin will remain HIGH unless data is being transmitted. Start bit B0 is always a HIGH-to-LOW transition, which tells the system that a new data word is about to be sent. Bits B1 through B8 are the data bits loaded into the transmitter on the sending end of the system. All eight bits of the maximum word length format are shown in the figure, even though truncated word lengths of 5, 6, or 7 bits are also allowable. The stop bit length can be programmed for 1, 1.5, or 2 bits, according to the needs of the system designer.

The number of data bits, the parity, and the number of stop bits are programmed onto the device using HIGH and LOW levels applied to certain pins designated for that purpose. For example, the WLS1 and WLS2 pins are used as *word length select* pins, and will set the data word length according to the following code system:

WORD LENGTH	WLS1	WLS2
5 bits	0	0
6 bits	1	0
7 bits	0	1
8 bits	1	1

Similarly, a two-bit stop code is selected by connecting SBS HIGH, but only when the data word is six, seven, or eight bits in length. If the data word is set to five bits length, which is used on Baudot teletypewriters, then the 1.5 bit stop code is used. If SBS is LOW, then the stop code is one bit in length. The parity is set by the EPE pin, and will be coded *odd* for a LOW and *even* for a HIGH.

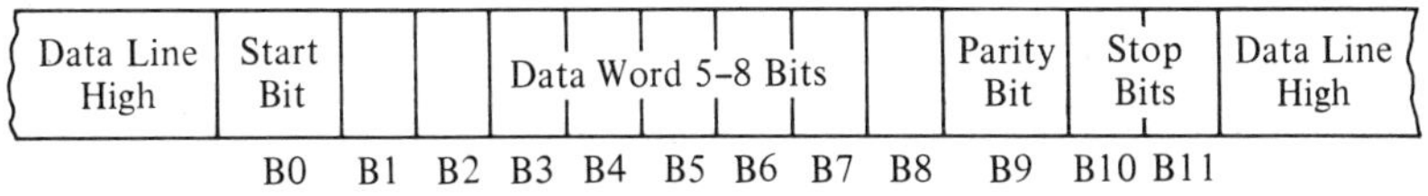

Data Line High	Start Bit	Data Word 5–8 Bits	Parity Bit	Stop Bits	Data Line High
	B0	B1 B2 B3 B4 B5 B6 B7 B8	B9	B10 B11	

FIGURE 10-9. UART data word format.

The clock on a UART system must be stable, so we cannot generally use RC timer-based clocks and expect proper performance, especially at higher baud rates. The frequency of the clock must be 16 times the baud rate. If we want to transmit data at 300 baud, for example, the oscillator frequency must be 300 × 16, or 4800 Hz. While this frequency is well within the range normally competent RC oscillators can produce, it is recommended that a crystal oscillator be used to ensure the stability and accuracy of the clock. An attractive alternative is the CMOS 4060 device, which contains an internal crystal or RC oscillator and a chain of binary divider stages.

The transmitter circuit is shown in Figure 10-10(a). Note that only the eight-bit input data, clock and serial output are required to make this circuit operational. The TRE, THRE and THRL signals are status flags, and are optional although they will probably be used in most practical cases. They convey information about the status of the information transfer, and these are sometimes needed in the software used to control the UART. A careful review of the meaning of each flag is necessary for those designers who wish to use the UART.

The basic receiver circuit for the UART is shown in Figure 10-10(b). We have a similar simplicity in the receiver section (one of the principle attractions held by LSI devices to equipment and instrument designers). Only the clock, serial input, and eight-bit parallel output lines are needed. Again, however, certain signals are available that will make some applications either easier or possible. These are the DR, OE, FE, and PE flags. You can look up the meanings of these signals, and those of the transmitter section, in Table 10-1.

Notice in the receiver section that we use an inverter from the *data received* terminal to reset the DRR terminal. This signal tells the UART to get ready for the next character and can be used to signal a distant transmitter that the UART is ready to receive another transmission.

One of the things about the UART that appeals to many designers is that the two sections (receiver and transmitter) can be used either independently or in a common system. In a simplex communications channel (one direction only), a transmitter-wired UART is used on the transmitter end, while a receiver-wired UART is used on the receiving end. In a half-duplex system (bidirectional communication, but only one direction at a time), both sections are used at each end, and the status flags can be used in a handshaking system to coordinate matters. Full-duplex operation is possible, but requires either a second channel (especially in radio links) or a second set of audio tones in hard-wired telephone line systems. Not all telephone lines, however, are amenable to full duplex operation, especially over long distance lines.

In dedicated instrument applications, the programming pins will

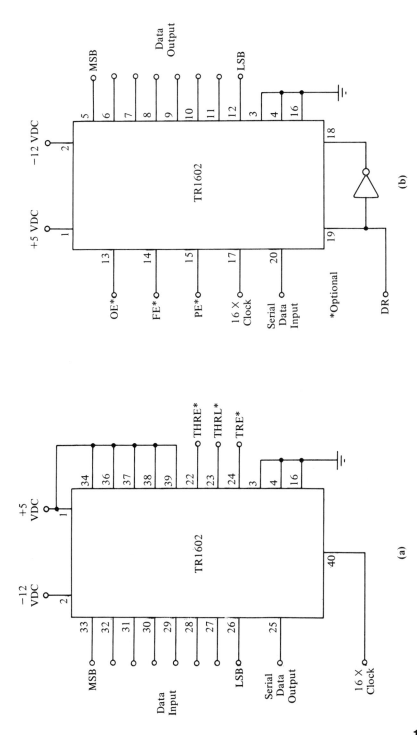

FIGURE 10-10. (a) TR1602UART—transmitter connections. (b) Receiver connections for the TR1602 UART.

TABLE 10-1

PIN NO.	MNEMONIC	FUNCTION
1	V_{CC}	+5 volts DC power supply.
2	V_{EE}	−12 volts DC power supply.
3	GND	Ground.
4	RRD	Receiver Register Disconnect. A high on this pin disconnects (i.e., places at high impedance) the receiver data output pins (5 through 12). A low on this pin connects the receiver data output lines to output pins 5 through 12.
5	RB_8	LSB ⎫
6	RB_7	⎪
7	RB_6	⎪
8	RB_5	⎬ Receiver data output lines
9	RB_4	⎪
10	RB_3	⎪
11	RB_2	⎪
12	RB_1	MSB ⎭
13	PE	Parity error. A high on this pin indicates that the parity of the received data does not match the parity programmed at pin 39.
14	FE	Framing Error. A high on this line indicates that no valid stop bits were received.
15	OE	Overrun Error. A high on this pin indicates that an overrun condition has occurred, which is defined as not having the DR flag (pin 19) reset before the next character is received by the internal receiver holding register.
16	SFD	Status Flag Disconnect. A high on this pin will disconnect (i.e., set to high impedance) the PE, FE, OE, DR, and THRE status flags. This feature allows the status flags from several UARTs to be bus-connected together.
17	RRC	16 × Receiver Clock. A clock signal is applied to this pin, and should have a frequency that is 16 times the desired baud rate (i.e., for 110 baud standard it is 16 × 110 baud, or 1760 hertz).
18	DRR	Data Receive Reset. Bringing this line low resets the data received (DR, pin 19) flag.
19	DR	Data Received. A high on this pin indicates that the entire character is received, and is in the receiver holding register.

Section 10.8 / SERIAL I/O PORTS

TABLE 10-1 *(Continued)*

PIN NO.	MNEMONIC	FUNCTION
20	RI	Receiver Serial Input. All serial input data bits are applied to this pin. Pin 20 must be forced high when no data is being received.
21	MR	Master Reset. A short pulse (i.e., a strobe pulse) applied to this pin will reset (i.e., force low) both receiver and transmitter registers, as well as the FE, OE, PE, and DRR flags. It also sets the TRO, THRE, and TRE flags (i.e., makes them high).
22	THRE	Transmitter Holding Register Empty. A high on this pin means that the data in the transmitter input buffer has been transferred to the transmitter register, and allows a new character to be loaded.
23	THRL	Transmitter Holding Register Load. A low applied to this pin enters the word applied to TB1 through TB8 (pins 26 through 33, respectively) into the transmitter holding register (THR). A positive-going level applied to this pin transfers the contents of the THR into the transmit register (TR), unless the TR is currently sending the previous word. When the transmission is finished the THR→TR transfer will take place automatically even if the pin 25 level transition is completed.
24	TRE	Transmit Register Empty. Remains high unless a transmission is taking place, in which case the TRE pin drops low.
25	TRO	Transmitter (Serial) Output. All data and control bits in the transmit register are output on this line. The TRO terminal stays high when no transmission is taking place, so the beginning of a transmission is always indicated by the first negative-going transition of the TRO terminal.
26	TB_8	LSB ⎫
27	TB_7	⎪
28	TB_6	⎪
29	TB_5	⎬ Transmitter input word.
30	TB_4	⎪
31	TB_3	⎪
32	TB_2	⎪
33	TB_1	MSB ⎭

TABLE 10-1 (*Continued*)

PIN NO.	MNEMONIC	FUNCTION
34	CRL	Control Register Load. Can be either wired permanently high, or be strobed with a positive-going pulse. It loads the programmed instructions (i.e., WLS1, WLS2, EPE, PI, and SBS) into the internal control register. Hard wiring of this terminal is preferred if these parameter never change, while switch or program control is preferred if the parameters do occasionally change.
35	PI	Parity Inhibit. A high on this pin disables parity generator/verification functions, and forces PE (pin 13) to a low logic condition.
36	SBS	Stop Bit(s) Select. Programs the number of stop bits that are added to the data word output. A high on SBS causes the UART to send two stop bits if the word length format is 6, 7, or 8 bits, and 1.5 stop bits if the 5-bit teletypewriter format is selected (on pins 37-38). A low on SBS causes the UART to generate only one stop bit.
37	WLS_1	Word Length Select. Selects character length, exclusive of parity bits, according to the rules given in the chart below:
38	WLS_2	

Word Length	WLS1	WLS2
5 bits	low	low
6 bits	high	low
7 bits	low	high
8 bits	high	high

PIN NO.	MNEMONIC	FUNCTION
39	EPE	Even Parity Enable. A high applied to this line selects even parity, while a low applied to this line selects odd parity.
40	TRC	16 × Transmit Clock. Apply a clock signal with a frequency that is equal to 16 times the desired baud rate. If the transmitter and receiver sections operate at the same speed (usually the case), then strap together TRC and RRC terminals so that the same clock serves both sections.

probably be hard-wired in the proper codes, but in many cases switches are used to allow the user to program as needed. You can also connect the UART control pins to an I/O port to permit programming of the UART under software control of the computer. This method, incidentally, is behind some of the programmable LSI I/O port devices that are

Section 10.8 / SERIAL I/O PORTS

designed to work with specific microcomputers. Such a system allows the user of the computer to "custom design" the communications port according to immediate program needs.

RS-232 Interfacing

The Electronic Industries Association (EIA) standard RS-232 pertains to a standardized serial data transmission scheme. The idea is to use the same connector (i.e., the DB-25 family), wired in the same manner all of the time, and to use the same voltage levels. Supposedly, one could connect together any two devices that provide RS-232 I/O without any problem (it usually works).

Modems, CTR terminals, printers (i.e., Model 43 Teletypewriter), and other devices will be fitted with RS-232 connectors. Some computers provide RS-232 I/O, and this feature can be added by using a set of Motorola ICs called RS-232 drivers/receivers. An RS-232 driver IC accepts TTL outputs from a computer or other device, and produces RS-232 voltage levels at its output. The RS-232 receiver does just the opposite. It takes RS-232 levels from the communications/interface and produces TTL outputs.

Unfortunately, the RS-232 is a very old standard, and it predates even the TTL standard. That is why it uses such odd voltage levels for logical-1 and logical-0.

Besides voltage levels, the standard also fixes the load impedances and the output impedances of the drivers.

There are actually two RS-232 standards—the older RS-232B and the current RS-232C (see Figure 10-11). In the older version, RS-232B, logical-1 is any potential in the -5- to -25-volt range, and logical-0 is anything between $+5$ and $+25$ volts. The voltages in the range -3 to $+3$ are a transition state, while $+3$ to $+5$ and -3 to -5 are undefined.

The speedier RS-232C standard narrows the limits to ± 15 volts. In addition, the standard fixes the load resistance to the range 3000 to 7000 ohms, and the driver output impedance that is low. The driver must provide a slew rate of 30 volts/microsecond. The Motorola MC1488 driver and MC1489 receiver ICs meet these specifications.

The standard wiring for the 25-pin DBM-25 connector used in RS-232 ports is shown in Table 10-2.

Current Loop Ports

The current loop port was designed specifically for use with teletypewriters, but it has been adopted over the years to a variety of com-

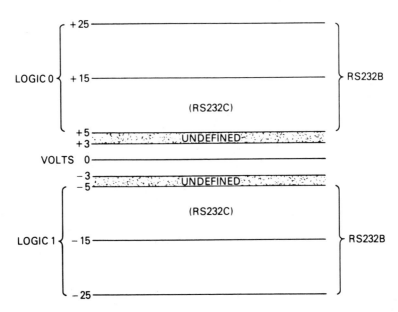

FIGURE 10-11. Diagram showing meaning of the RS-232 EIA standards.

munications problems in digital instruments. The original 60 (and later 20) milliampere current loop systems were intended for Baudot Teletype® machines, and were used to energize the solenoids in the printer. But the same idea has also been adopted for use with a variety of printers other than Teletypes® and is also found in certain other instruments that must communicate with computers. The 60 mA version of the current loop is obsolete but is included here because it is often necessary to design into older, existing systems.

Figure 10-12(a) shows the most basic circuit for a 60-mA machine. An external 130-volt dc power supply is needed. The current loop circuit consists of the dc supply, resistor R2, the teletypewriter machine, and *c-e* path of transistor Q1.

Diode D1 is used as a spike suppressor. The solenoid coils will produce a spike-like pulse (i.e., high amplitude, short duration) every time the current flow in one of the coils is interrupted. Diode D1 is connected to suppress these spikes, and is used mainly to protect transistor Q1.

Transistor Q1 can be any high-voltage power transistor that is capable of handling a 60-mA collector current. Q1 acts as a switch to turn the loop on and off.

If a HIGH appears on the LSB of the selected output port, then Q1 is forward-biased. Its *c-e* path conducts current, closing the loop. When the LSB of the output port is LOW, then Q1 is reverse-biased. Under this condition, its *c-e* path is turned off, so the loop is open.

TABLE 10-2
EIA RS-232 Pin-outs for Standard DB-25 connecta

PIN NO.	RS232 NAME	FUNCTION
1	AA	Chassis ground
2	BA	Data from terminal
3	BB	Data received from modem
4	CA	Request to send
5	CB	Clear to send
6	CC	Data set ready
7	AB	Signal ground
8	CF	Carrier detection
9	undef	
10	undef	
11	undef	
12	undef	
13	undef	
14	undef	
15	DB	Transmitted bit clock, internal
16	undef	
17	DD	Received bit clock
18	undef	
19	undef	
20	CD	Data terminal ready
21	undef	
22	CE	Ring indicator
23	undef	
24	DA	Transmitted bit clock, external
25	undef	

It is best to adjust resistor R2 to obtain a loop current of 60 mA. Place a HIGH on the LSB of the selected port, and press one of the teletypewriter keys. A millammeter placed at the point indicated in Figure 10-12(a) will show the current. Adjust the resistor (R2) for a flow of 60 mA.

It is probably best if all high-voltage circuits are isolated from your computer's output. A fault in transistor Q1 could otherwise cause damage

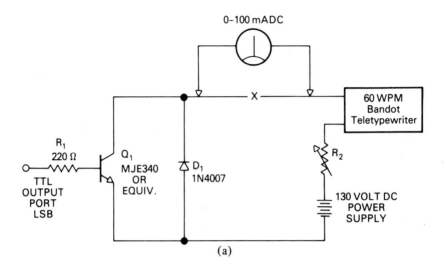

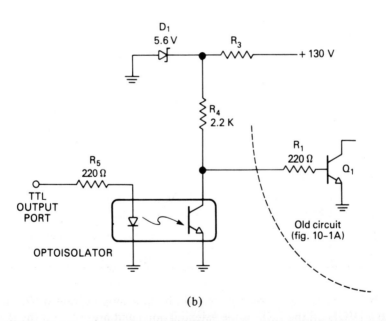

FIGURE 10-12. (a) Simple circuit to interface old-style Baudot teletypewriters. Adjust R2 for 60 mA in the loop. (b) Circuit above modified to isolate the teletypewriter from the computer output circuitry.

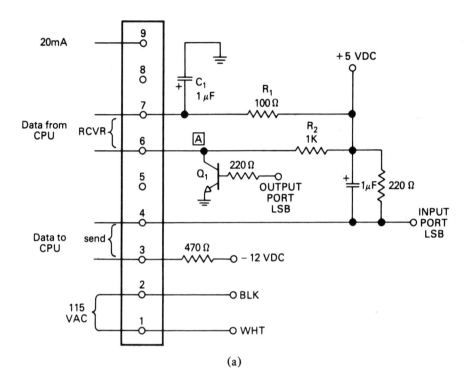

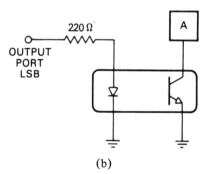

FIGURE 10-13. (a) Circuit to connect computer output port to the Model 33 teletypewriter. Terminal block shown is found under the top cover of the Mdl. 33, on the right rear when viewed from the front of the keyboard. Use separate +5 volt dc power supplies for best results. (b) Modification of the standard circuit to allow isolation of the computer from the teletypewriter.

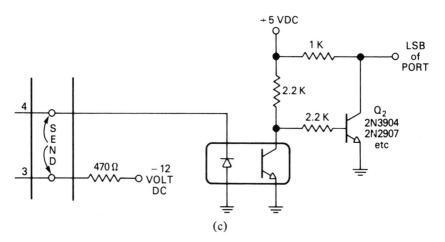

FIGURE 10-13. (c) Different circuit to accomplish the same job.

to the output port circuits. An appropriate circuit for this is shown in Figure 10-12(b). The secret is to use an optoisolator device. On the computer side of the device is an LED, while on the teletypewriter side is an optotransistor. The transistor will be turned off unless the LED is turned on. The collector of the optoisolator transistor is connected to the point in the previous circuit that connected to the computer. This collector is also connected to a 5.6-volt dc power supply that is derived from the +130-volt dc power supply used in the current loop. On the computer side, the LED is connected, through a current-limiting resistor (R5), to the LSB of the selected port.

When the LSB of the output port is HIGH, then the LED is turned on. This turns on the transistor in the optoisolator, shorting out the bias to the current loop transistor. This action turns off the loop. Similarly, the LOW on the LSB of the port turns off the transistor, so Q1 is turned on, closing the loop. The action in this circuit is inverted, so it is necessary to complement the Z80 accumulator before outputting data. Alternatively, you could use one other transistor inverter, between the isolator and Q1, to invert the output of the isolator.

Figure 10-13 shows a circuit that is used to interface the model 33 to a Z80 output port. Looking from the front panel, there is a terminal strip on the right-rear side of the Model 33. This terminal strip, shown schematically in Figure 10-13(b), contains the send/receive connections for the teletypewriter.

The *receive* side of the machine (terminals 6 and 7) contains the loop, so that the solenoids can be pulled in. The *send* side is merely a set of contact closures. In my own experience, this circuit has produced

Section 10.8 / SERIAL I/O PORTS

some problems. If the loop is turned on after the microcomputer is loaded and ready to work, a random pulse seems to change a few (important) bits in a few memory locations. The problem is partially relieved by using +5-volt and −12-volt power supplies that are completely divorced from the computer power supply. But I like the approach shown in Figures 10-13(b) and 10-13(c). We would use R1, R2, and C1 [from Figure 10-13(a)], but replace Q1 with the transistor from the optoisolator (connect the collector to point "A"). The LED is connected, again through a current-limiting resistor, to the LSB of the selected output port.

We can use the −12-volt supply to drive the LED, or the +5-volt supply (in which case, the polarity is reversed). The isolator transistor (Q1) drives an inverter stage (Q2). When the LED is turned on, Q2 is turned off, so the LSB of the selected input port is HIGH. But if the LED is off, then Q2 is turned on, dropping the LSB of the input port to zero.

11
Signals and Noise

All electronic instruments deal with *signals* of one sort or another; typical signals may be bioelectric potentials (as in electrocardiographic and other physiological instruments), a voltage or current from a pressure, flow, or temperature transducer, or a digital signal from something like a shaft encoder or computer. Signals, then, are very important to the design of electronic instrumentation of all sorts. In fact, many electronic instruments, especially the digital types, are little more than dedicated signals processor circuits. We don't have the space in this text to dig deeply into the matter of signals. For those who want a deeper treatment, let me recommend Dr. Frank Stanley's book, *Digital Signal Processing*[1] as a good introductory primer.

11.1 DIFFERENT TYPES OF SIGNALS

The electronic signals that we must deal with are basically *voltage or current functions of time*. We find at least three different types of signal within this definition: *continuous, sampled,* and *digitized*. Figure 11-1(a) shows the basic continuous signal. Although a straight line form is shown here for sake of simplicity, it could be a continuous voltage function of time. The one characteristic important to this discussion is that the signal can be measured at any instant of time from T_a to T_b, and can take on any value from Z_a to Z_b. In other words, the signal in Figure 11-1(a) is continuous in both range and domain. This is the typical form of signals used in most analog instruments.

A *sampled* signal is shown in Figure 11-1(b). In this case, the range

---

[1] Stanley, F.; *Digital Signal Processing,* Reston Publishing Co., 11480 Sunset Hills Rd., Reston, Va., 22090.

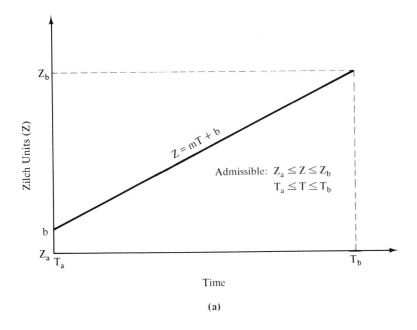

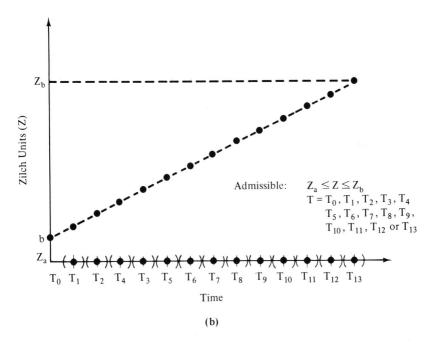

FIGURE 11-1. (a) Function that is continuous in both range and domain. (b) Function that has a continuous range and discrete domain.

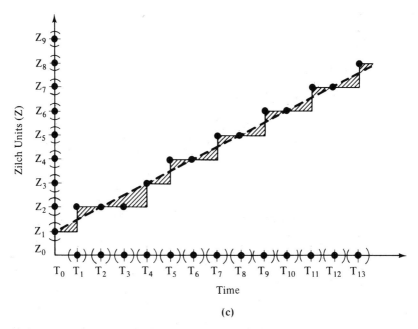

(c)

FIGURE 11-1(c). Discrete range and domain.

(Zilch units!?!) can take on any value from Z_a to Z_b, while the time units can exist only at certain specific values, i.e., T_0, T_1, T_2, etc. This signal is an example of a continuous range, discrete domain signal. In a typical electronic instrument that uses such a signal there will be a sample and hold circuit that will take the value at a predetermined time.

The *digitized* signal is shown in Figure 11-1(c). In this signal both the range and domain can exist only in certain discrete values. This is the typical form of the signal that is used in a digital computer. The computer clock will determine the value for time, while the nature of the N-bit binary word used in the computer will determine the admissible values of Z. Note that the dotted line represents the function of Figure 11-1(a), while the solid line represents the digitized value at various instants in time. The shaded area between the dotted and solid lines gives an indication of the error between the correct value and the digitized value.

The size of the steps between values of Z depends upon the number of bits used to make the data word. For example, if we have an eight-bit A/D converter, then the full-scale range can be represented by 256 discrete steps. Of course, the larger the number of bits, the more resolution

Section 11.1 / DIFFERENT TYPES OF SIGNALS

is possible in any given application. Be wary, however, of using large size A/D converters unless (1) you need the extra bits and, (2) the rest of the circuit can support the large size data converter.

All signals that are continuous and periodic can be represented by a series consisting of sine and cosine terms and their coefficients. This series is called the Fourier Series by mathematicians. There is an elementary rule for sampled and digitized signals that the sampling rate (samples per second) must be at least *twice* the highest significant frequency in the Fourier series. In practice, it is often found that better results obtain when a sampling rate of four to six times the highest frequency is selected.

Figure 11-2 shows how the sampling rate might affect the reproducibility of a sampled and/or digitized signal. The signal in Figure 11-2(a) is shown in sampled form in Figure 11-2(b). The sample is held at its present value for each instant in time and will not change until the next sample is taken. Obviously, the more samples, the nearer this signal is to the original.

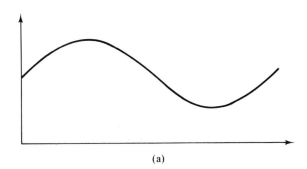

(a)

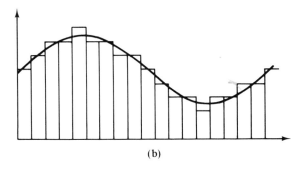

(b)

FIGURE 11-2. Sampled signal.

11.2 NOISE

"Noise" in an electronic instrument is any signal that is not wanted. In medical and scientific instruments such signals are often called artifact. For the rigorous definition of noise we must consider spurious signals, variations in the signal that are not caused by the measured phenomenon, crud added to the signal by the instrument, the digitization/sampling error terms discussed above, and any other addition to the signal. Of prime consideration when designing microcomputer-based instruments, however, is a particular form of noise that emanates from the AC power mains: high voltage transients and "suck-outs." The high voltage transient may be due to lightning (even far away) or to high inductance industrial machinery. In a factory or hospital there are many devices that will throw high voltage spikes (see Figure 11-3) onto the AC power mains. In a hospital setting, for example, there are multi-kilowatt X-ray machines that will throw one or more HV spikes onto the power line every time an X-ray is taken. Large AC motors, air conditioning compressor clutches, and other sources will also produce these spikes because of the high inductance of the devices.

Lightning need not be either local or a direct hit to the power line in order to be a problem for microcomputers. The HV transient will travel over a great distance to become a problem. The HV can be *induced* into the line from any nearby lightning strike or from an aerial burst overhead. There are several solutions to this problem, including the use of an MOV device (General Electric) in the primary circuit of the power supply for the computer, and isolating the power supply from the AC mains with a noise suppression transformer such as the Topaz Electronics (3855 Ruffin Road, San Diego, Ca. 92123) model shown in Figure 11-4.

The problems created in the computer by high voltage transients include alteration of memory location and register contents, jumps during

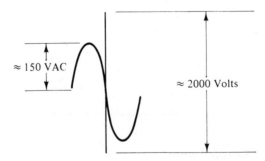

FIGURE 11-3. Transient spike on 110 volt (RMS) AC power line.

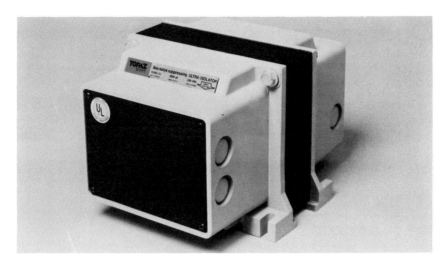

FIGURE 11-4. Transient suppression transformer isolates digital equipment from noisy power line. (*Courtesy of Topaz Electronics.*)

the execution of software to locations that are unpredictable, and even destruction of the electronics. Obviously, the engineer who is designing such an instrument must take into consideration the effects of AC power mains transient HV voltage pulses.

The "suck-out" problem is the momentary reduction of amplitude of the AC power mains voltage. This problem is caused by heavy-current devices starting up. You can see this in your home—when the air conditioner or clothes dryer starts up there is a surge current that will momentarily dim the lights. If a computer is connected to such a power line, then it is likely to experience a problem. One such problem is retriggering (at an inappropriate time) of the power-on reset circuit. The cure for this type of problem is to use very heavy filter capacitors in the DC power supply. The charge stored in the filters will be dumped into the circuit during the short duration suck-out and that will often prevent the problems. This phenomenon is not to be confused with brown-outs, which are voltage reductions by the power company. These must be overcome by using a constant voltage AC line conditioner (also see Topaz Electronics Catalog).

12
Operational Amplifiers

12.1 OBJECTIVES

1. To learn basic operational amplifier theory.
2. To learn the major operational amplifier configurations used in instrumentation circuits.
3. To learn to derive the elementary operational amplifier transfer functions.
4. To learn how to use laboratory amplifiers.

12.2 SELF-EVALUATION QUESTIONS

Before studying the material in this chapter, try answering the questions given below. These questions test your knowledge of the subject matter. If you cannot answer any particular question, then look for the answer as you read the text.

1. Write the transfer equation for (a) *inverting follower*, (b) *noninverting follower*.
2. Draw a circuit for an active *integrator* using an operational amplifier.
3. How would you cure *offset voltage* problems?
4. Write the transfer equation for an active operational amplifier *differentiator*.
5. What should be the value of a *compensation resistor*, R_c?

12.3 OPERATIONAL AMPLIFIERS: AN INTRODUCTION

The operational amplifier has been in existence for several decades, but only in the last 10 or 15 years has it come into its own as an almost universal electronic building block. The term *operational* is derived from the fact that these devices were originally designed for use in analog computers to solve *mathematical operations*. The range of circuit applications today, however, has increased immensely, so the operational amplifier has survived and prospered, even though analog computers, in which they were once a principal constituent, are now almost in eclipse.

Keep in mind, however, that even though the programmable analog computer is no longer used extensively, many instruments are little more than a nonprogrammable, dedicated-to-one-chore analog computer with a numeric readout of some sort.

In this chapter we will examine the *gross,* or large-scale, properties of the basic operational amplifier and learn to derive the transfer equations for most common operational amplifier circuits using only Ohm's law, Kirchoff's law, and the basic properties of all operational amplifiers.

One of the profound beauties of the modern, integrated-circuit, operational amplifier is its simplicity when viewed from the outside world. Of course, the inner workings are complex, but they are of little interest in our discussion of the operational amplifier's gross properties. We will limit our discussion somewhat by considering the operational amplifier as a *black box,* and that allows for a very simple analysis in which we relate the performance to the universal transfer function for all electronic circuits, namely, E_{out}/E_{in}.

12.4 PROPERTIES OF THE IDEAL OP-AMP

An *ideal* operational amplifier is a *gain block,* or black box, that has the following general properties:

1. Infinite *open-loop* (i.e., no feedback) gain ($A_{vol} = \infty$)
2. Infinite input impedance ($Z_{in} = \infty$)
3. Zero output impedance ($Z_0 = \infty$)
4. Infinite bandwidth ($f_0 = \infty$)
5. Zero noise generation

Of course, it is not possible to obtain a real IC operational amplifier that meets these properties—they are *ideal*—but if we read "infinite" as "very, very high," and "zero" as "very, very low," then the approximations of the ideal situation are very accurate. Real IC operational amplifiers, for example, can have an open-loop voltage gain from 50,000 to over 1,000,000 so it can be classed as *relatively infinite,* so the equations work in most cases.

12.5 DIFFERENTIAL INPUTS

Figure 12-1 shows the basic symbol for the common operational amplifier, including power terminals. In many schematics of operational amplifier circuits, the V_{cc} and V_{ee} power terminals are deleted, so the drawing will be less "busy."

Note that there are two input terminals, labeled (−) and (+). The terminal labeled (−) is the *inverting* input. The output signal will be out of phase with signals applied to this input terminal; i.e., there will be a 180-degree phase shift. The terminal labeled (+) is the *noninverting* input, so output signals will be in phase with signals applied to this input. It is important to remember that these inputs look into *equal* open-loop gains, so they will have equal but opposite effects on the output voltage.

At this point let us add one further property to our list of ideal properties:

6. Differential inputs follow each other.

This property implies that the two inputs will behave as if they were at the same potential, especially under static conditions. In Figure 12-2 we see an inverting follower circuit in which the noninverting (+) input is grounded. The sixth property allows us, in fact requires us, to treat the inverting (−) input as if it were *also* grounded. Many textbooks and magazine articles like to call this phenomenon a "virtual" ground, but that is a term that serves only to confuse the reader. It is better to accept as a basic

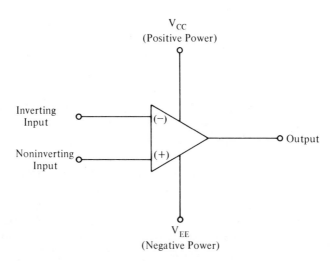

FIGURE 12-1. Symbol for an operational amplifier.

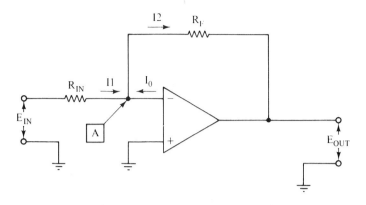

FIGURE 12-2. Inverting follower.

axiom of operational amplifier circuitry that, for purposes of calculation and voltage measurement, the (−) input will be grounded if the (+) input is actually grounded.

12.6 ANALYSIS USING KIRCHOFF AND OHM

We know from Kirchoff's current law that the algebraic sum of all currents entering and leaving a point in a circuit must be zero. The total current flow into and out of point A in Figure 12-2, then, must be *zero*. Three possible currents exist at this point: input current $I1$, feedback current $I2$, and any currents flowing into or out of the (−) input terminal of the operational amplifier, I_0. But according to ideal property 2, the input impedance of this type of device is infinite. Ohm's law tells us that by

$$I_0 = E/Z_{in} \qquad (12\text{-}1)$$

current I_0 is zero, because E/Z_{in} is zero. So, if current I_0 is equal to zero, we conclude that $I1 + I2 = 0$ (Kirchoff's law). Since this is true, then

$$I2 = -I1 \qquad (12\text{-}2)$$

We also know that

$$I1 = E_{in}/R_{in} \qquad (12\text{-}3)$$

and $$I2 = E_{out}/R_f \qquad (12\text{-}4)$$

By substituting Equations (12-3) and (12-4) into Equation (12-2), we obtain the result

$$\frac{E_{out}}{R_f} = \frac{-E_{in}}{R_{in}} \qquad (12\text{-}5)$$

Solving for E_{out} gives us the transfer function normally given in operational amplifier literature for an inverting amplifier:

$$E_{out} = -E_{in} \times \frac{R_f}{R_{in}} \qquad (12\text{-}6)$$

Example 12-1

Calculate the output voltage from an inverting operational amplifier circuit if the input signal is 100 mV, the feedback resistor is 100 kΩ, and the input resistor is 10 kΩ.

Solution

$$E_{out} = E_{in} \times (R_f/R_{in}) \qquad (12\text{-}6)$$
$$E_{out} = (0.1 \text{ V})(10^5 \text{ }\Omega/10^4 \text{ }\Omega)$$
$$E_{out} = (0.1 \text{ V})(10) = \mathbf{1 \text{ V}}$$

The term R_f/R_{in} is the voltage gain factor, and is usually designated by the symbol A_v, which is written as

$$A_v = -R_f/R_{in} \qquad (12\text{-}7)$$

We sometimes encounter Equation (12-6) written using the left-hand side of Equation (12-7):

$$E_{out} = -A_v E_{in} \qquad (12\text{-}8)$$

When designing simple inverting followers using operational amplifiers, use Equations (12-7) and (12-8). Let us look at a specific example. Suppose that we have a requirement for an amplifier with a gain of 50. We want to drive this amplifier from a source that has an output impedance of 1000 ohms. A standard rule of thumb for designers to follow is to make the input impedance not less than 10 times the source impedance, so in this case the amplifier must have a source impedance that is equal to or greater than 10,000 ohms (10 kΩ). This requirement sets the value of the input re-

Section 12.7 / NONINVERTING FOLLOWERS

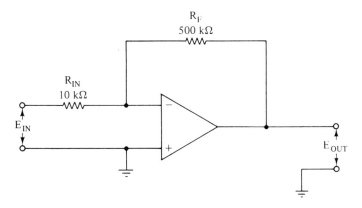

FIGURE 12-3. Gain-of-50 inverting follower.

sistor at 10 kΩ or higher, but in this example we select a 10-kΩ value for R_{in}.

$$A_v = R_f/R_{in} \tag{12-9}$$

$$50 = R_f/10{,}000 \text{ ohms} \tag{12-10}$$

$$R_f = 500{,}000 \text{ ohms}$$

Our gain-of-50 amplifiers will look like Figure 12-3.

12.7 NONINVERTING FOLLOWERS

The inverting follower circuits of Figures 12-2 and 12-3 suffer badly from low input impedance, especially at higher gains, because the input impedance is the value of R_{in}. This problem becomes especially acute when we attempt to obtain even moderately high gain figures from low-cost devices. Although some types of operational amplifier allow the use of 500-kΩ to 2-megohm input resistors, they are costly and often uneconomical. The *noninverting follower* of Figure 12-4 solves this problem by using the input impedance problem very nicely, because the input impedance of the op-amp is typically very, very high (ideal property 2).

We may once again resort to Kirchoff's law to derive the transfer equation from our basic ideal properties. By property 6 we know that the inputs tend to follow each other, so the inverting input can be treated as if it were at the same potential as the noninverting input, which is E_{in}, the input signal voltage. We know that

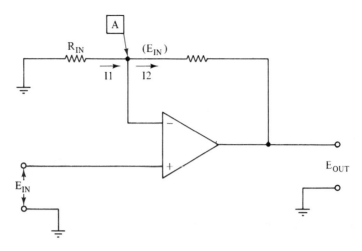

FIGURE 12-4. Noninverting follower with gain.

$$I1 = I2 \tag{12-11}$$

$$I1 = E_{in}/R_{in} \tag{12-12}$$

$$I2 = (E_{out} - E_{in})/R_f \tag{12-13}$$

By substituting Equations (12-12) and (12-13) into Equation (12-11) we obtain

$$I1 = I2 \tag{12-14}$$

$$\frac{E_{in}}{R_{in}} = \frac{E_{out} - E_{in}}{R_f} \tag{12-15}$$

Solving Equation (12-15) for E_{out} results in the transfer equation for the noninverting follower amplifier circuit.

Multiply both sides by R_f $\quad \dfrac{R_f E_{in}}{R_{in}} = E_{out} - E_{in} \tag{12-16}$

Add E_{in} to both sides $\quad \dfrac{R_f E_{in}}{R_{in}} + E_{in} = E_{out} \tag{12-17}$

Factor out E_{in} $\quad E_{in} \times \left[\dfrac{R_f}{R_{in}} + 1\right] = E_{out} \tag{12-18}$

Section 12.8 / OPERATIONAL AMPLIFIER POWER SUPPLIES

Example 12-2

Calculate the output voltage for 100 mV (i.e., 0.1 V) input in a noninverting follower amplifier if R_f is 100 kΩ and R_{in} is 10 kΩ.

Solution

$$E_{out} = E_{in}[(R_f/R_{in}) + 1] \qquad (12\text{-}18)$$

$$E_{out} = (0.1 \text{ V})[(10^5 \text{ Ω}/10^4 \text{ Ω}) + 1]$$

$$E_{out} = (0.1 \text{ V})(10 + 1)$$

$$E_{out} = (0.1 \text{ V})(11) = \mathbf{1.1 \text{ V}}$$

In this discussion we have arrived at both of the transfer functions commonly used in operational amplifier design, using only the basic properties, Ohm's law, and Kirchoff's current law. We may safely assume that the operational amplifier is merely a feedback device that generates a current that exactly cancels the input current. Figure 12-5 gives a synopsis of the characteristics of the most popular operational amplifier configurations. The unity gain noninverting following of Figure 12-5(c) is a special case of the circuit in Figure 12-5(b), in which $R_f/R_{in} = 0$. In this case, the transfer equation becomes

$$E_{out} = E_{in}(0 + 1) \qquad (12\text{-}19)$$

$$E_{out} = E_{in}(1) \qquad (12\text{-}20)$$

$$E_{out} = E_{in} \qquad (12\text{-}21)$$

12.8 OPERATIONAL AMPLIFIER POWER SUPPLIES

Although almost every circuit using operational amplifiers uses a dual polarity power supply, it is possible to operate the device with a single polarity supply. An example of single supply operation might be in equipment designed for mobile operation, or in circuits where the other circuitry requires only a single polarity supply, and an op-amp or two are but minority features in the design. It is, however, generally better to use the bipolar supplies as intended by the manufacturer.

There are two separate power terminals on the typical operational amplifier device, and these are marked V_{cc} and V_{ee}. The V_{cc} supply is con-

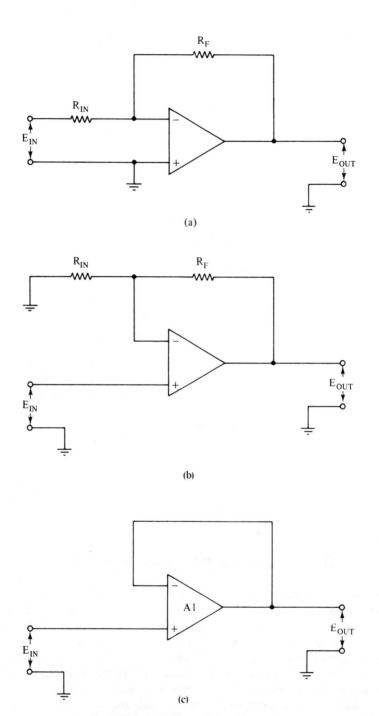

FIGURE 12-5. (a) Inverting follower. (b) Noninverting gain follower. (c) Unity gain noninverting follower.

Section 12.9 / PRACTICAL DEVICES: SOME PROBLEMS

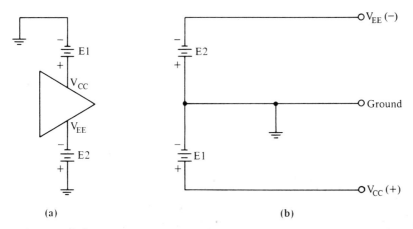

FIGURE 12-6. (a) Power supply requirements of an operational amplifier. (b) Typical operational amplifier configuration.

nected to a power supply that is *positive* to ground, while the V_{ee} supply is *negative* with respect to ground. These supplies are shown in Figure 12-6. Keep in mind that, although batteries are shown in the example, regular power supplies may be used instead. Typical values for V_{cc} and V_{ee} range from ±3 volts dc to ±22 volts dc. In many cases, perhaps most, the value selected for these potentials will be between ±9 volts dc and ±15 volts dc.

There is one further constraint placed on the operational amplifier power supply: $V_{cc} - V_{ee}$ must be less than some specified voltage, usually 30 volts. So, if V_{cc} is +18 volts dc, then V_{ee} must not be greater than (30 − 18), or 12 volts dc.

12.9 PRACTICAL DEVICES: SOME PROBLEMS

Before we can properly apply operational amplifiers in real equipment we must learn some of the limitations of real-world devices. The devices that we have considered up until now have been *ideal,* so they do not exist. Real IC operational amplifiers carry price tags of less than half a dollar up to several dozen dollars each. The lower the cost, generally, the less ideal the device.

Three main problems exist in real operational amplifiers: offset current, offset voltage, and frequency response. Of less importance in many cases is noise generation.

In real operational amplifier devices the input impedance is less than infinite, and this implies that a small input bias current exists. The input current may flow into or out of the input terminals of the operational amplifier. In other words, current I_0 of Figure 12-2 is *not* zero, so it will produce an output voltage equal to $-I_0 \times R_f$. The cure for this problem is shown in Figure 12-7, and involves placing a *compensation resistor* between the noninverting input terminal and ground. This tactic works because the currents in the respective inputs are approximately equal. Since resistor R_c is equal to the parallel combination of R_f and R_{in}, it will generate the same *voltage drop* that appears at the inverting input. The resultant output voltage, then, is zero, because the two inputs have equal but opposite polarity effect on the output.

Output offset voltage is the value of E_{out} that will exist if the input end of R_{in} is grounded (i.e., $E_{in} = 0$). In the ideal device, E_{out} would be zero under this condition, but in real devices there may be some offset potential present. This output potential can be forced to zero by any of the circuits in Figure 12-8.

The circuit in Figure 12-8(a) uses a pair of *offset null* terminals found on many, but not all, operational amplifiers. Although many IC operational amplifiers use this technique, some do not. Alternatively, the offset range may be insufficient in some cases, In either event, we may use the circuit of Figure 12-8(b) to solve the problem.

The offset null circuit of Figure 12-8(b) creates a current flowing in resistor $R1$ to the summing junction of the operational amplifier. Since the offset current may flow either *into* or *out of* the input terminal, the null control circuit must be able to supply currents of both polarities. Because of this requirement, the ends of the potentiometer ($R1$) are connected to V_{cc} and V_{ee}.

In many cases, it is found that the offset is small compared with nor-

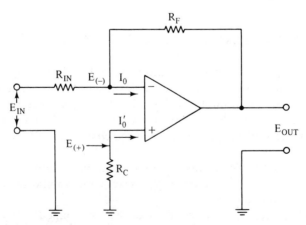

FIGURE 12-7. Use of a compensation resistor.

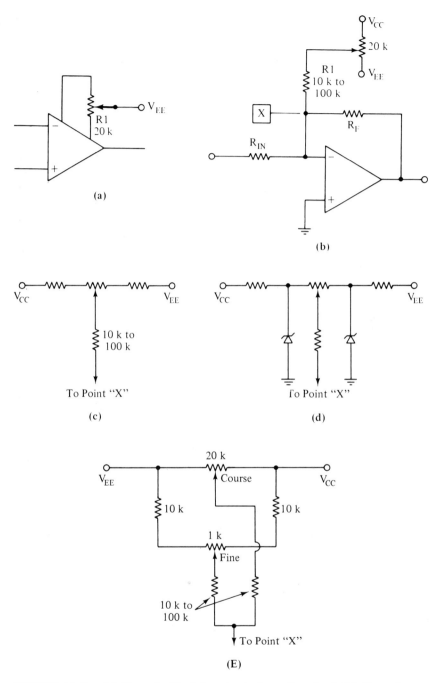

FIGURE 12-8. (a) Use of offset terminals to null output. (b) Use of summing current to null output. (c)-(e) High resolution offset null circuits.

mally expected values of input signal voltage. This is especially true in low-gain applications, in which case the nominal offset current will create such a low output error that no action need be taken. In still other cases, the offset of each stage in a cascade chain of amplifiers may be small, but their cumulative effect may be a large offset error. In this type of situation, it is usually sufficient to null only one of the stages late in the chain (i.e., close to the output stage).

In those circuits where the offset is small, but critical, it may be useful to replace $R1$ and $R2$ of Figure 12-8(b) with one of the resistor networks of Figures 12-8(c) through 12-8(e). These perform essentially the same function, but have superior resolution. That is to say, there is a smaller change in output voltage for a single turn of the potentiometer. This type of circuit will have a superior resolution in any event, but even further improvement is possible if a 10-turn (or more) potentiometer is used.

12.10 DC DIFFERENTIAL AMPLIFIERS

The fact that an IC operational amplifier has two complementary inputs, inverting and noninverting, makes it a natural for application as a *differential amplifier*. These circuits produce an output voltage that is proportional to the *difference* between two ground-referenced input voltages. Recall from our previous discussion that the two inputs of an operational amplifier have equal but opposite effect tn the output voltage. If the same voltage or two equal voltages are applied to the two inputs (i.e., a *common mode* voltage, $E3$ in Figure 12-1), then the output voltage will be zero. The transfer equation for a differential amplifier is

$$E_{out} = A_v(E1 - E2) \qquad (12\text{-}22)$$

So, if $E1 = E2$, then $E_{out} = 0$.

The circuit of Figure 12-9 shows a simple differential amplifier using a single IC operational amplifier. The voltage gain of this circuit is given by

$$A_v = R3/R1 \qquad (12\text{-}23)$$

provided that $R1 = R2$ and $R3 = R4$.

The main appeal of this circuit is that it is economical, requiring but one IC operational amplifier. It will reject common mode voltages reasonably well if the equal resistors are well matched. A glaring problem exists, however, and that is a low input impedance. Additionally, with the

Section 12.10 / DC DIFFERENTIAL AMPLIFIERS

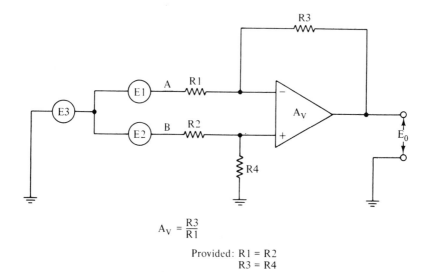

FIGURE 12-9. Differential amplifier with input voltages.

problems existing in real operational amplifiers, this circuit may be a little difficult to tame in high-gain applications. As a result, designers frequently use an alternate circuit in these cases.

In recent years, the instrumentation amplifier (IA) of Figure 12-10 has become popular because it alleviates most of the problems associated with the circuit of Figure 12-9. The input stages are noninverting followers, so they will have a characteristically high input impedance. Typical values run to as much as 1000 megohms.

The instrumentation amplifier is relatively tolerant of different resistor ratios used to create voltage gain. In the simplest case, the differential voltage gain is given by

$$A_v = \frac{2R3}{R1} + 1 \qquad (12\text{-}24)$$

provided that $R3 = R2$, and $R4 = R5 = R6 = R7$.

It is interesting to note that the common-mode rejection ratio is not seriously degraded by mismatch of resistors $R2$ and $R3$; only the gain is affected. If these resistors are mismatched, then a differential voltage gain error will be introduced.

The situation created by Equation (12-24) results in having the gain of $A3$ equal to unity (i.e., 1), and that is a waste. If gain in $A3$ is desired, then Equation (12-24) must be rewritten into the form

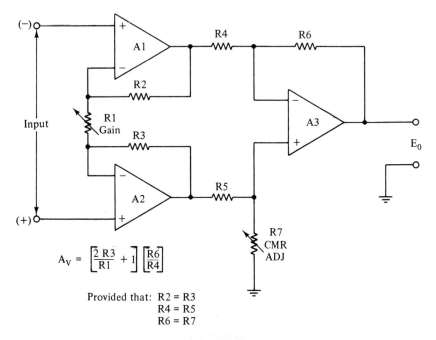

FIGURE 12-10. Instrumentation amplifier.

$$A_v = \left[\frac{2R3}{R1} + 1\right]\left[\frac{R7}{R5}\right] \quad (12\text{-}25)$$

Example 12-3

Calculate the differential voltage gain of an instrumentation amplifier that uses the following resistor values: $R3 = 33$ kΩ, $R1 = 2.2$ kΩ, $R5 = 3.3$ kΩ, and $R7 = 15$ kΩ.

Solution

$$A_v = [(2R3/R1) + 1)(R7/R5)] \quad (12\text{-}25)$$

$$A_v = \left[\frac{(2)(33 \text{ k}\Omega)}{(2.2 \text{ k}\Omega)} + 1\right]\left[\frac{(15 \text{ k}\Omega)}{(3.3 \text{ k}\Omega)}\right]$$

$$A_v = 141$$

One further equation that may be of interest is the general expression from which the other instrumentation amplifier transfer equations are derived:

$$A_v = \frac{R7(R1 + R2 + R3)}{R1R6} \qquad (12\text{-}26)$$

which remains valid provided that the ratio $R7/R6 = R5/R4$.

Equation (12-26) is especially nice, since you need not be concerned with matched pairs of precision resistors, but only that their ratios be equal.

12.11 PRACTICAL CIRCUIT

In this section we will consider a practical design example using the instrumentation amplifier circuit. The particular problem required a frequency response to 100 kHz, and shielded input lines. But the latter requirement would also deteriorate the signal at high frequencies because of the shunt capacitance of the input cables. To overcome this problem a *high-frequency compensation* control is built into the amplifier. Voltage gain is approximately 10.

The circuit to the preamplifier is shown in Figure 12-11. It is, of course, the instrumentation amplifier of Figure 12-10 with some modifications. When the frequency response is less than 10 kHz or so, we may use any of the 741-family devices (i.e., 741, 747, 1456, and 1458), but premium performance demands a better operational amplifier. In this case, one of the most economical is the RCA CA3140, although an L156 would also suffice.

Common mode rejection can be adjusted to compensate for any mismatch in the resistors or IC devices by adjusting $R10$. This potentiometer is adjusted for zero output when the same signal is applied simultaneously to both inputs.

The frequency response characteristics of this preamplifier are shown in Figures 12-12 through 12-16. The input in each case was a 1000-Hz square wave from a function generator. The waveform in Figure 12-12 shows the output signal when resistor $R9$ is set with its wiper closest to ground. Notice that it is essentially square, showing only a small amount of roll-off of high frequencies. The waveform in Figure 12-13 is the same signal when $R9$ is at maximum resistance. This creates a small amount of regenerative (i.e., positive) feedback; although it is not sufficient to start oscillation, it will enhance amplification of high frequencies.

The problem of oscillation can be quite serious (Figure 12-15), however, if certain precautions are not taken, most of which involve limiting

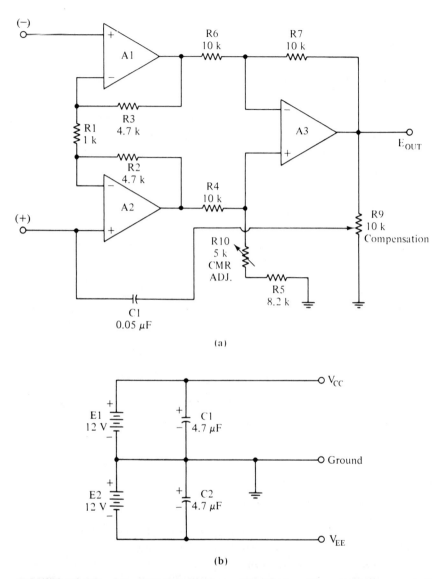

FIGURE 12-11. (a) Instrumentation amplifier with capacitance null. (b) Power supply for (a).

the amplitude of the feedback signal. This goal is realized by using a 2200-ohm resistor in series with the potentiometer.

Another source of oscillation is the value of $C1$. When a 0.001-μF capacitor is used at $C1$, an 80-kHz oscillation is created (see Figure

Section 12.11 / PRACTICAL CIRCUIT

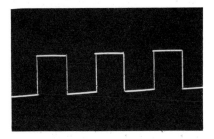

FIGURE 12-12. Square wave.

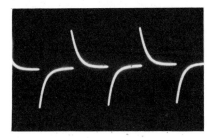

FIGURE 12-13. Differentiated square wave.

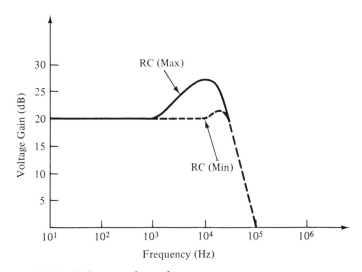

FIGURE 12-14. Voltage gain-vs-frequency.

FIGURE 12-15. Ringing on square wave.

12-16). The frequency response is shown in Figure 12-14. To obtain any particular response curve, modify the values of $C1$ and $R9$.

12.12 DIFFERENTIAL AMPLIFIER APPLICATIONS

Differential amplifiers find application in many different instrumentation situations. Of course, it should be realized that they are required wherever a differential signal voltage is found. Less obvious, perhaps, is that they are used to acquire signals or to operate in control systems, in the presence of large noise signals. Many medical applications, for example, use the differential amplifier, because they look for minute biopotentials in the presence of strong 60-Hz fields from the ac power mains.

Another class of applications is the amplification of the output signal from a Wheatstone bridge, and this is shown in Figure 12-17. If one side of the bridge's excitation potential is grounded, then the output voltage is a differential signal voltage. This signal can be applied to the inputs of a differential amplifier or instrumentation amplifier to create an amplified, single-ended, output voltage.

A "rear end" stage suitable for many operational amplifier instrumentation projects is shown in Figure 12-18. This circuit consists of three

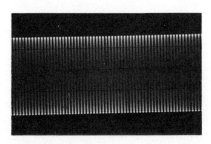

FIGURE 12-16. Eighty-kHz oscillation.

Section 12.12 / DIFFERENTIAL AMPLIFIER APPLICATIONS 151

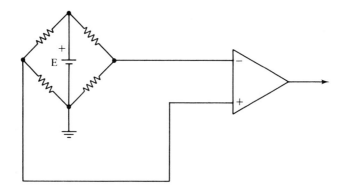

FIGURE 12-17. Differential amplifier used to amplify output of Wheatstone bridge.

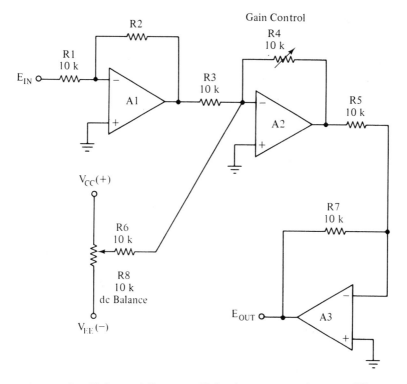

FIGURE 12-18. Universal "rear end" for instrumentation amplifiers and other purposes.

low-cost operational amplifier ICs. Since they follow most of the circuit gain, we may use low-cost devices such as the 741 in this circuit. The gain of this circuit is given by $R2/10^4$.

12.13 INTEGRATORS

Figure 12-19 shows the basic operational amplifier *integrator* circuit. The transfer equation for this circuit may be derived in the same manner as before, with due consideration for $C1$.

$$I2 = -I1 \tag{12-27}$$

but
$$I1 = E_{in}/R1 \tag{12-18}$$

and
$$I2 = C1(dE_0/dt) \tag{12-19}$$

Substituting Equations (12-28) and (12-29) into Equation (12-27) results in

$$\frac{C1\ dE_0}{dt} = \frac{-E_{in}}{R1} \tag{12-30}$$

We may now solve Equation (12-30) for E_0 by integrating both sides.

$$\int \frac{C1\ dE_0}{dt}\ dt = -\int \frac{E_{in}}{R1}\ dt \tag{12-31}$$

$$C1E_0 = \frac{-1}{R1} \int_0^t E_{in}\ dt \tag{12-32}$$

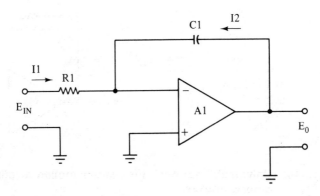

FIGURE 12-19. Integrator circuit.

Section 12.14 / DIFFERENTIATORS

$$E_0 = \frac{-1}{R1C1} \int_0^t E_{in}\, dt \tag{12-33}$$

Equation (12-33), then, is the transfer equation for the operational amplifier integrator circuit.

Example 12-4

A constant potential of 2 volts is applied to the input of the integrator in Fig. 12-19 for 3 seconds. Find the output potential if $R1 = 1$ megohm and $C1 = 0.5\ \mu F$.

Solution

$$E_0 = \frac{-1}{R1C1} \int_0^t E_{in}\, dt \tag{12-33}$$

$$E_0 = \frac{-E_{in}}{R1C1} \int_0^3 dt$$

$$E_0 = \frac{(-2\ \text{V})(t)}{(10^6\ \Omega)(5 \times 10^{27}\ \text{F})} \Big|_0^3$$

$$E_0 = \frac{(-2\ \text{V})(3\ \text{s})}{(5 \times 10^{22}\ \text{s})} - 0 = -12\ \textbf{volts}$$

Note that the *gain* of the integrator is given by the term $1/R1C1$. If small values of $R1$ and $C1$ are used, then the gain can be very large. For example, if $R1 = 100\ k\Omega$, and $C1 = 0.001\ \mu F$, then the gain is 10,000. A very small input voltage in that case will saturate the output very quickly. In general, the *time constant* $R1C1$ should be longer than the period of the input waveform.

12.14 DIFFERENTIATORS

An operational amplifier *differentiator* is formed by reversing the roles of $R1$ and $C1$ in the integrator, as shown in Figure 12-20. We know that

$$I2 = -I1 \tag{12-34}$$

$$I1 = \frac{C1\ dE_{in}}{dt} \tag{12-35}$$

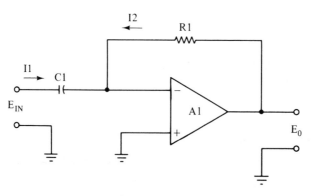

FIGURE 12-20. Differentiator circuit.

and $$I2 = E_0/R1 \tag{12-36}$$

Substituting Equations (12-35) and (12-36) into Equation (12-34) results in

$$\frac{E_0}{R1} = \frac{-C1\ dE_{in}}{dt} \tag{12-37}$$

Solving Equation (12-37) for E_0 gives us the transfer equation for an operational amplifier differentiator circuit:

$$E_0 = -R1C1\frac{dE_{in}}{dt} \tag{12-38}$$

Example 12-5

A 12-volt/second ramp function voltage is applied to the input of an operational amplifier differentiator, in which $R1 = 1$ megohm and $C1 = 0.2\ \mu F$. What is the output voltage?

Solution

$$E_0 = -R1C1(dE_{in}/dt) \tag{12-38}$$
$$E_0 = -(10^6\ \Omega)(2 \times 10^{-7}\ F)(12\ V/s)$$
$$E_0 = -(2 \times 10^{-1}\ s)(12\ V/s) = -2.4\ V$$

The differentiator time constant $R1C1$ should be set very short relative to the period of the waveform being differentiated, or in the case of square waves, triangle waves, and certain other signals, the time constant should be short compared with the *rise time* of the leading edge.

12.15 LOGARITHMIC AND ANTILOG AMPLIFIERS

Figure 12-21(a) shows an elementary *logarithmic amplifier* circuit using a bipolar transistor in the feedback loop. We know that the collector current bears a logarithmic relationship to the base-emitter potential, V_{be}:

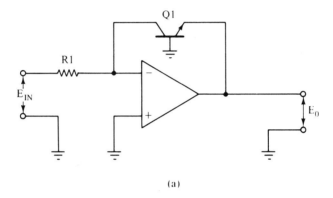

(a)

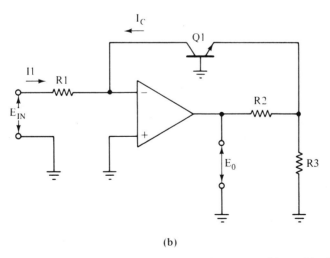

(b)

FIGURE 12-21. (a) Logarithmic amplifier. (b) Improved logarithmic amplifier.

$$V_{be} = \frac{KT}{q} \ln\left[\frac{I_c}{I_s}\right] \qquad (12\text{-}39)$$

where V_{be} = the base-emitter potential in volts (V)

K = Boltzmann's constant 1.38×10^{-23} joules/degree Kelvin (j/°K)

T = the temperature in degrees Kelvin (°K)

q = the electronic charge (1.6×10^{-19} coulombs)

I_c = the collector current in amperes (A)

I_s = the reverse saturation current for the transistor in amperes (A)

At 27°C (300°K), i.e., room temperature, the term KT/q evaluates to approximately 26 mV (i.e., 0.026 V), so Equation (12-39) becomes

$$V_{be} = 26 \text{ mV } \ln(I_c/I_s) \qquad (12\text{-}40)$$

But $V_{be} = E_0$, and $I_c = E_{in}/R1$, so we may safely say that

$$E_0 = 26 \text{ mV } \ln\left[\frac{E_{in}}{I_s R1}\right] \qquad (12\text{-}41)$$

But I_s is a constant if the temperature is also constant, and $R1$ is constant under all conditions, so, by the rule that the logarithm of a constant is also a constant, we may state that Equation (12-41) is the transfer function for the natural logarithmic amplifier. For base-10 logarithms:

$$E_0 = 60 \text{ mV } \log_{10}\left[\frac{E_{in}}{I_s R1}\right] \qquad (12\text{-}42)$$

The relationship of Equations (12-41) and (12-42) allows us to construct amplifiers with logarithmic properties. If $Q1$ is in the feedback loop of an operational amplifier, then the output voltage E_0 will be proportional to the logarithm of input voltage E_{in}. If, on the other hand, the transistor is connected in series with the input of the operational amplifier (see Figure 12-22), then the circuit becomes an antilog amplifier.

Both of these circuits exhibit a strong dependence on temperature, as evidenced by the T term in Equation (12-39). In actual practice, then, some form of temperature correction must be used. Two forms of temperature correction are commonly used: *compensation* and *stabilization*.

The compensation method uses temperature-dependent resistors, i.e., *thermistors*, to regulate the gain of the circuit with changes in temper-

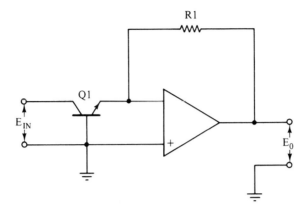

FIGURE 12-22. Antilog amplifier.

ature. For example, it is common practice to make $R3$ in Figure 12-21(b) a thermistor.

The stabilization method requires that the temperature of $Q1$, and preferably the op-amp also, be held constant. In the past, this has meant that the components must be kept inside an electrically heated oven, but today other techniques are used. One manufacturer builds a temperature-controlled hybrid logarithmic amplifier by nesting the op-amp and transistor on the same substrate as a class-A amplifier. Such an amplifier, under zero-signal conditions, dissipates very nearly constant heat. After the chip comes to equilibrium, the temperature will remain constant.

In the case of the antilog amplifier:

$$I_c = E_0/R1 \qquad (12\text{-}43)$$

and
$$E_{in} = V_{be} \qquad (12\text{-}44)$$

$$E_{in} = 26 \text{ mV ln } (E_0/R1I_s) \qquad (12\text{-}45)$$

12.16 SUMMARY

1. The transfer function of an operational amplifier circuit can be determined by the basic properties, Ohm's law, and Kirchoff's law.
2. Operational amplifiers can be used as amplifiers, integrators, differentiators, and so forth, to provide numerous functions, both linear and nonlinear.
3. Operational amplifiers can be configured to provide differential inputs.
4. The gain of an operational amplifier circuit can be set by the ratio of two resistors.

12.17 RECAPITULATION

Now go back and try answering the questions at the beginning of the chapter. When you are finished, answer the questions and work the problems given below. Place a mark beside any problem or question that you cannot answer, and then go back to the text and reread appropriate sections.

12.18 QUESTIONS

1. Write the transfer function for an inverting follower.
2. Write the transfer function for a noninverting follower.
3. Write the transfer function for a differential amplifier.
4. Write the transfer function for an instrumentation amplifier.
5. Write the transfer function for an operational amplifier integrator.
6. Write the transfer function for an operational amplifier differentiator.
7. State four of the basic properties of an ideal operational amplifier.
8. Describe in your own words "virtual ground." State the property of ideal operational amplifiers on which this term is based.
9. An operational amplifier is designed to operate from two power supplies. V_{cc} is _____ with respect to ground, while V_{ee} is _____ with respect to ground.
10. Name three mechanisms by which an output offset voltage is created.
11. List two different methods for eliminating the output offset voltage.
12. The common mode voltage gain for a perfect operational amplifier is _____.
13. The gain of an operational amplifier integrator is given by the expression: _____.
14. Write the transfer equations for (a) base-e, (b) base-10 logarithmic amplifiers.
15. What steps may be taken to reduce the effect of temperature on the logarithmic amplifier?

12.19 PROBLEMS

1. Calculate the voltage gain of an inverting follower if the feedback resistor is 560 kΩ, and the input resistor is 82 kΩ.
2. Calculate the voltage gain of an inverting follower if the feedback resistor is 100 kΩ and the input resistor is 10 kΩ.
3. Calculate the voltage gain of an inverting follower if the feedback resistor is 5.6 kΩ and the input resistor is 560 ohms.

Section 12.19 / PROBLEMS

4. Calculate the voltage gain of an inverting follower if the feedback resistor is 5 kΩ and the input resistor is 10 kΩ.
5. Calculate the voltage gain of a noninverting follower if the feedback resistor is 120 kΩ and the input resistor is 56 kΩ.
6. Calculate the voltage gain of a noninverting follower if the feedback resistor is 10 kΩ and the input resistor is 10 kΩ.
7. Calculate the voltage gain of a noninverting follower if the feedback resistor is 0 ohms and the input resistor is 100 kΩ.
8. Calculate the voltage gain of a noninverting follower if the feedback resistor is 2.2 kΩ and the input resistor is 4.7 kΩ.

In Problems 9 through 11 refer to Figure 12-23.

9. In Figure 12-23 point A is at a potential of +1 volt, point B is at +6 volts, and point D is at 0 volts. Find the output voltage if $R2 = 10$ kΩ and $R1 = 9.1$ kΩ.
10. In Figure 12-23 point A is grounded, point B is at a dc potential of −3 volts, and point D is at a potential of +1.5 volts dc. Find the output potential if $R1 = R2 = 100$ kΩ, and $R3 = R1/2$.

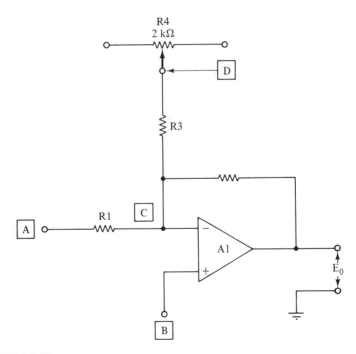

FIGURE 12-23.

11. In Figure 12-29 points A and B are grounded, and point D is at a potential of -1 volt dc. Find $R2$ if $R3 = 2.2$ kΩ and $E_0 = +10$ volts.
12. Find the gain of an instrumentation amplifier such as Figure 12-10 if $R1 = 500$ ohms, $R2 = 16$ kΩ, $R4 = 3.9$ kΩ, and $R7 = 27$ kΩ.
13. An operational amplifier integrator uses a 0.69-μF capacitor and an 820-kΩ resistor. What will the output voltage be if a 100-mV dc level is applied to the input for 4 seconds?
14. What is the gain of an integrator that uses a 0.1-μF capacitor and a 47-kΩ resistor?
15. An operational amplifier differentiator uses a 1-megohm resistor and a 0.68-μF capacitor. What output voltage exists when an 8 V/sec ramp function is applied to the input?
16. Find V_{be} for a transistor operated at 40°C if the reverse saturation current at that temperature is 10^{-12} amperes, and the collector current is 1.2 mA.
17. A logarithmic amplifier is adjusted for the natural log operation mode. Find the output voltage if the reverse saturation current is 10^{-13} amperes, the input resistor is 100 kΩ, and $E_{in} = 100$ mV.
18. What is the *slope factor* of a natural logarithmic amplifier at 37°C?
19. A voltage-to-current converter consists of an operational amplifier and a 100-kΩ resistor in the negative feedback loop. Find E_0 if the input current is 20 μA.
20. Derive the transfer function for an antilog amplifier. Do not refer back to the text.

12.20 REFERENCES

1. Edward Bannon, *Operational Amplifiers: Theory and Servicing.* Reston, Va.: Reston Publishing Co., 1975.
2. Arpad Barna, *Operational Amplifiers.* New York: Wiley-Interscience, 1971.
3. Joseph J. Carr, *Op-Amp Circuit Design & Applications.* Blue Ridge Summit, Pa.: TAB Books, 1976.
4. Jerald G. Graeme, Gene Tobey, and L.P. Huelsman, eds., *Operational Amplifiers—Design and Applications.* New York: McGraw-Hill Book Company, 1973.
5. Jerald G. Graeme, *Applications of Operational Amplifiers—Third Generation Techniques.* New York: McGraw-Hill Book Company, 1973.
6. Walter Jung, *IC Op-Amp Cookbook.* Indianapolis: Howard W. Sams & Co., 1974.
7. David F. Stout and Milton Kaufman, eds., *Handbook of Operational Circuit Design.* New York: McGraw-Hill Book Company, 1976.

13
Transducers

13.1 OBJECTIVES

1. To learn the principles behind the operation of common types of transducer.
2. To learn how to specify and apply transducers.
3. To learn the limitations of certain transducers.
4. To learn about certain application problems and how they are solved.

13.2 SELF-EVALUATION QUESTIONS

Before studying the material in this chapter, try answering the questions given below. These questions test your knowledge of the subject matter. If you cannot answer any particular question, then look for the answer as you read the text.

1. Describe the *differences* between *bonded* and *unbonded* piezoresistive strain gages.
2. List three different types of temperature transducer.
3. List two types of gas flow transducer.
4. Define "transducer" in your own words.
5. How may a force be measured if only displacement transducers are available?

13.3 TRANSDUCERS AND TRANSDUCTION

Not all of the physical variables lend themselves to direct input into electronic instruments and circuits. Unfortunately, electronic circuits operate only with inputs that are either currents or voltages. So when one is measuring nonelectrical physical quantities it becomes necessary to provide a device that converts physical parameters such as *force, displacement, temperature, etc.*, into proportional voltages or currents. The transducer is such a device.

> *Definition:* A *transducer* is a device or apparatus that converts nonelectrical physical parameters into electrical signals, i.e., currents or voltages, that are proportional to the value of the physical parameter being measured.

Transducers take many forms, and may be based on a wide variety of physical phenomena. Even when one is measuring the *same* parameter, different instruments may use different types of transducer.

This chapter will not be an exhaustive catalogue treatment covering all transducers—the manufacturer's data sheets may be used for that purpose—but we will discuss some of the more common *types* of transducer used in scientific, industrial, medical, and engineering applications.

13.4 THE WHEATSTONE BRIDGE

Many forms of transducer create a variation in an electrical resistance, inductance, or capacitance in response to some physical parameter. These transducers are often in the form of a Wheatstone bridge, or one of the related ac bridge circuits. In many cases where the transducer itself is not in the form of a bridge, it is used in a bridge circuit with other components forming the other arms of the bridge.

13.5 STRAIN GAGES

All electrical conductors possess some amount of electrical resistance. A bar or wire made of such a conductor will have an electrical resistance that is given by

$$R = \rho(L/A) \qquad (13\text{-}1)$$

where R = the resistance in ohms (Ω)

ρ = the *resistivity constant*, a property specific to the conductor material, given in units of ohm-centimeters (Ω-cm)

L = the length in centimeters (cm)

A = the cross-sectional area in square centimeters (cm^2)

Section 13.5 / STRAIN GAGES

Example 13-1

A constantan (i.e., 55 percent copper, 45 percent nickel) round wire is 10 cm long and has a radius of 0.01 mm. Find the electrical resistance in ohms. (*Hint:* The resistivity of constantan is 44.2×10^{-6} Ω-cm).

Solution

$$R = \rho(L/A) \tag{13-1}$$

$$R = \frac{(44.2 \times 10^{-6} \text{ ohm-cm})(10 \text{ cm})}{\pi \left[0.01 \text{ mm} \times \frac{1 \text{ cm}}{10 \text{ mm}} \right]^2}$$

$$R = \frac{(4.42 \times 10^{-4} \text{ ohm-cm}^2)}{\pi (0.001 \text{ cm})^2}$$

$$R = \frac{(4.42 \times 10^{-4} \text{ ohm-cm}^2)}{\pi 10^{-6} \text{ cm}^2} = \textbf{141 ohms}$$

Note that the resistivity factor (ρ) in Equation (13-1) is a constant, so if length L or area A can be made to vary under the influence of an outside parameter, then the electrical resistance of the wire will change. This phenomenon is called *piezoresistivity*, and is an example of a transducible property of a material.

Definition: Piezoresistivity is the change in the electrical resistance of a conductor due to changes in length and cross-sectional area. In piezoresistive materials mechanical deformation of the material produces changes in electrical resistance.

Figure 13-1 shows how an electrical conductor can use the piezoresistivity property to measure *strain*, i.e., *forces* applied to it in *compression* or *tension*. In Figure 13-1(a) we have a conductor at rest, in which no forces are acting. The length is given as L_0 and the cross-sectional area as A_0. The resistance of this conductor, from Equation (13-1), is

$$R_0 = \rho(L_0/A_0) \tag{13-2}$$

where ρ = the resistivity as defined previously
 R_0 = the resistance in ohms (Ω) when no forces are applied
 L_0 = the resting, i.e., no force, length in cm
 A_0 = the resting cross-sectional area in cm^2

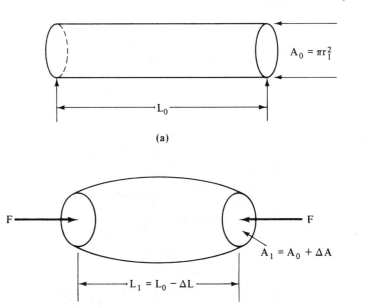

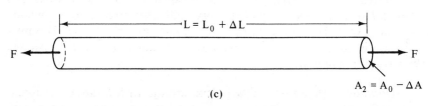

FIGURE 13-1. (a) Unstrained metal bar. (b) Metal bar in compression. (c) Metal bar in tension.

But in Figure 13-1(b) we see the situation where a compression force of magnitude F is applied along the axis in the inward direction. The conductor will deform, causing the length L_1 to decrease to $(L_0 - \Delta L)$, and the cross-sectional area to increase to $(A_0 + \Delta A)$. The electrical resistance decreases to $(R_0 - \Delta R)$:

$$R_1 = (R_0 - \Delta R) \propto \frac{(L_0 - \Delta L)}{(A_0 + \Delta A)} \qquad (13\text{-}3)$$

Similarly, when a tension force of the same magnitude (i.e., F) is applied—i.e., a force that is directed outward along the axis—the length increases to $(L_0 + \Delta L)$, and the cross-sectional area decreases to

Section 13.5 / STRAIN GAGES

$(A_0 - \Delta A)$. The resistance will increase to

$$R_2 = (R_0 - \Delta R) \propto \frac{(L_0 + \Delta L)}{(A_0 + \Delta A)} \tag{13-4}$$

The *sensitivity* of the strain gage is expressed in terms of unit change of electrical resistance for a unit change in length, and is given in the form of a *gage factor S:*

$$S = \frac{(\Delta R/R)}{(\Delta L/L)} \tag{13-5}$$

where S = the gage factor (dimensionless)
R = the unstrained resistance of the conductor
ΔR = the change in resistance due to strain
L = the unstrained length of the conductor
ΔL = the change in length due to strain

Example 13-2

Find the gage factor of a 128-ohm conductor that is 24 mm long, if the resistance changes 13.3 ohms and the length changes 1.6 mm under a tension force.

Solution

$$S = (\Delta R/R)/(\Delta L/L) \tag{13-5}$$

$$S = \frac{(13.3 \; \Omega/128 \; \Omega)}{(1.6 \; \text{mm}/24 \; \text{mm})}$$

$$S = \frac{1.04 \times 10^{-1}}{6.67 \times 10^{-2}} = \mathbf{1.56}$$

We may also express the gage factor in terms of the length and diameter of the conductor. Recall that the diameter is related to the cross-sectional area (i.e., $A = \pi d^2/4 = \pi r^2$), so the relationship between the gage factor S and these other factors is given by

$$S = 1 + 2\frac{(\Delta d/d)}{(\Delta L/L)} \tag{13-6}$$

Example 13-3

Calculate the gage factor S if a 1.5-mm-diameter conductor that is 24 mm long changes length by 1 mm and diameter by 0.02 mm under a compression force.

Solution

$$S = 1 + 2\frac{(\Delta d/d)}{(\Delta d/d)} \tag{13-6}$$

$$S = 1 + \frac{2[(0.02 \text{ mm})/(1.5 \text{ mm})]}{(1 \text{ mm})/(24 \text{ mm})}$$

$$S = 1 + \frac{(2)(1.3 \times 10^{-2})}{(4.2 \times 10^{-2})}$$

$$S = 1 + (2)(0.31) = \mathbf{1.62}$$

Note that the expression $(\Delta L/L)$ is sometimes denoted by the Greek letter ϵ, so Equations (13-5) and (13-6) become

$$S = 1 + \frac{2(\Delta d/d)}{\epsilon}$$

$$S = \frac{(\Delta R/R)}{\epsilon}$$

Gage factors for various metals vary considerably. Constantan, for example, has a gage factor of approximately 2, while certain other common alloys have gage factors between 1 and 2. At least one alloy (92 percent platinum, 8 percent tungsten) has a gage factor of 4. Semiconductor materials such as germanium and silicon can be doped with impurities to provide custom gage factors between 50 and 250. The problem with semiconductor strain gages, however, is that they exhibit a marked sensitivity to temperature changes. Where semiconductor strain gages are used, either a thermally controlled environment or temperature compensating circuitry must be provided.

13.6 BONDED AND UNBONDED STRAIN GAGES

Strain gages can be classified as *unbonded* or *bonded*. The categories refer to the method of construction used. Figure 13-2 shows both methods of construction.

Section 13.6 / BONDED AND UNBONDED STRAIN GAGES

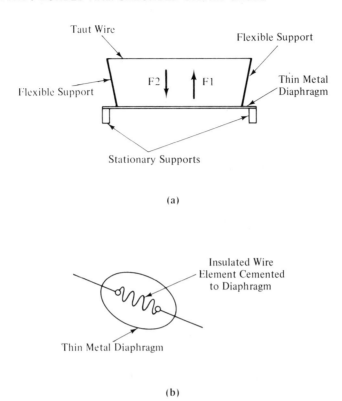

FIGURE 13-2. (a) Unbonded strain gage. (b) Bonded strain gage.

The unbonded type of strain gage is shown in Figure 13-2(a), and consists of a wire resistance element stretched taut between two flexible supports. These supports are configured in such a way as to place tension or compression forces on the taut wire when external forces are applied. In the particular example shown, the supports are mounted on a thin metal diaphragm that flexes when a force is applied. Force $F1$ will cause the flexible supports to spread apart, placing a tension force on the wire and increasing its resistance. Alternatively, when force $F2$ is applied, the ends of the flexible supports tend to move closer together, effectively placing a compression force on the wire element, thereby reducing its resistance. In actuality, the wire's resting condition is *tautness*, which implies a tension force. So $F1$ increases the tension force from normal, and $F2$ decreases the normal tension.

The bonded strain gage is shown in Figure 13-2(b). In this type of device a wire or semiconductor element is cemented to a thin metal diaphragm. When the diaphragm is flexed, the element deforms to produce a resistance change.

The linearity of both types can be quite good, provided that the elastic limits of the diaphragm and the element are not exceeded. It is also necessary to insure that the ΔL term is only a very small percentage of L.

In the past it has been "standard wisdom"—i.e., the opinions of those who make purchasing decisions—that bonded strain gages are more rugged, but less linear, than unbonded models. Although this may have been true at one time, recent experience has shown that modern manufacturing techniques produce linear, reliable instruments of both types.

13.7 STRAIN GAGE CIRCUITRY

Before a strain gage can be useful, it must be connected into a circuit that will convert its resistance changes to a current or voltage output. Most applications are voltage output circuits.

Figure 13-3(a) shows the *half-bridge* (so called because it is actually half of a Wheatstone bridge circuit) or *voltage divider* circuit. The strain gage element of resistance R is placed in series with a fixed resistance $R1$ across a stable and well-regulated voltage source E. The output voltage E_0 is found from the voltage divider equation

$$E_0 = \frac{ER}{R + R1} \qquad (13\text{-}7)$$

Equation (13-7) describes the output voltage E_0 when the transducer is at rest, i.e., nothing is stimulating the strain gage element. When the element is stimulated, however, its resistance changes a small amount ΔR. To simplify our discussion we will adopt the standard convention used in many texts of letting $h = \Delta R$.

$$E_0 = \frac{E(R + h)}{(R \pm h) + R1} \qquad (13\text{-}8)$$

Another half-bridge is shown in Figure 13-3(b), but in this case the strain gage is in series with a *constant current source* (CCS), which will maintain current I at a constant level regardless of changes in strain gage resistance. The normal output voltage E_0 is

$$E_0 = IR \qquad (13\text{-}9)$$

for nonstimulated conditions, and

Section 13.7 / STRAIN GAGE CIRCUITRY

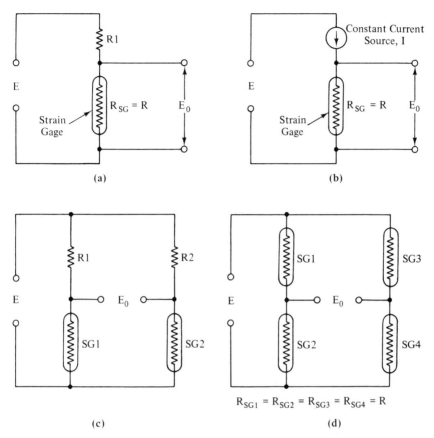

FIGURE 13-3. (a) Constant voltage strain gage circuit (half-bridge type).
(b) Constant current strain gage circuit (half-bridge type).
(c) Two strain-gage elements in Wheatstone bridge.
(d) Four-active-element strain gage Wheatstone bridge.

$$E_0 = I(R \pm h) \qquad (13\text{-}10)$$

under stimulated conditions.

The half-bridge circuits suffer from one major defect: Output voltage E_0 will always be present regardless of the stimulus. Ideally, in any transducer system, we want E_0 to be zero when the stimulus is also zero, and take a value proportional to the stimulus when the stimulus value is nonzero. A Wheatstone bridge circuit in which one or more strain gage elements form the bridge arms has this property.

Figure 13-3(c) shows a circuit in which strain gage elements $SG1$ and $SG2$ form two bridge arms and fixed resistors $R1$ and $R2$ form the other two arms. It is usually the case that *SG*1 and *SG*2 will be configured so that their actions oppose each other; that is, under stimulus, *SG*1 will have a resistance $R + h$ and *SG*2 will have a resistance $R - h$, or vice versa.

One of the most linear forms of transducer bridge is the circuit of Figure 13-3(d) in which all four bridge arms contain strain gage elements. In most such transducers all four strain gage elements have the same resistance, i.e., R, which has a value between 100 ohms and 1000 ohms in most cases.

Recall that the output voltage from a Wheatstone bridge is the difference between the voltages across the two half-bridge dividers. The following equations hold true for bridges in which one, two, or four equal active elements are used.

One active element:

$$E_0 = \frac{E}{4}\left[\frac{h}{R}\right] \tag{13-11}$$

(accurate ± 5 percent, provided that $h \leq 0.1$)

Two active elements:

$$E_0 = \frac{E}{2}\left[\frac{h}{R}\right] \tag{13-12}$$

Four active elements:

$$E_0 = \frac{Eh}{R} \tag{13-13}$$

where, for all three equations,

E_0 = the output potential in volts (V)
E = the excitation potential in volts (V)
R = the resistance of all bridge arms
h = the quantity ΔR, the change in resistance of a bridge arm under stimulus

(These equations apply only when all the bridge arms have equal resistances under zero stimulus conditions.)

Example 13-4

A transducer that measures force has a nominal resting resistance of 300 ohms and is excited by +7.5 volts dc. When a 980-dyne force is applied, all four equal-resistance bridge elements change resistance by 5.2 ohms. Find the output voltage E_0.

Solution

$$E_0 = E(h/R) \tag{13-13}$$
$$E_0 = (7.5 \text{ V})(5.2 \text{ }\Omega/300 \text{ }\Omega)$$
$$E_0 = (7.5 \text{ V})(5.2)/(300) = \mathbf{0.13 \text{ V}}$$

13.8 TRANSDUCER SENSITIVITY (ψ)

When designing electronic instrumentation systems involving strain gage transducers, it is convenient to use the *sensitivity factor* (denoted by the Greek letter "psi" — ψ), which relates the output voltage in terms of the excitation voltage and the applied stimulus. In most cases, we see a specification giving the number of microvolts or millivolts output per volt of excitation potential per unit of applied stimulus (i.e., $\mu V/V/Q_0$ or $mV/V/Q_0$).

$$\psi = E'_0/V/Q_0 \tag{13-14a}$$

and

$$\psi = \frac{E'_0}{V \times Q_0} \tag{13-14b}$$

where E'_0 = the output potential
 V = one unit of potential, i.e., 1 volt
 Q_0 = one unit of stimulus

The sensitivity is often given as a specification by the transducer manufacturer. From it we can predict output voltage for any level of stimulus and excitation potential. The output voltage, then, is found from

$$E_0 = \psi E Q \tag{13-15}$$

where E_0 = the output potential in volts (V)
 ψ = the sensitivity in $\mu V/V/Q_0$

E = the excitation potential in volts (V)
Q = the stimulus parameter

Example 13-5

A well-known medical arterial blood pressure transducer uses a four-element piezoresistive Wheatstone bridge with a sensitivity of 5 microvolts per volt of excitation per Torr of pressure, i.e., 5 $\mu V/V/T$ (*Note:* 1 Torr = 1 mm Hg). Find the output voltage if the bridge is excited by 5 volts dc and 120 Torr of pressure is applied.

Solution

$$E_0 = \psi E Q \qquad (13\text{-}15)$$

$$E_0 = \frac{5 \mu V}{V \times T} \times (5 \text{ V}) \times (120 \text{ T})$$

$$E_0 = (5 \times 5 \times 120) \mu V = \mathbf{3000\ \mu V}$$

13.9 BALANCING AND CALIBRATING THE BRIDGE

Few, if any, Wheatstone bridge strain gages meet the ideal condition in which all four arms have exactly equal resistances. In fact, the bridge resistance specified by the manufacturer is a *nominal* value only. There will inevitably be an *offset voltage*, i.e., $E_0 \neq 0$ when $Q = 0$. Figure 13-4 shows a circuit that will balance the bridge when the stimulus is zero. Potentiometer $R1$, usually a type with 10 or more turns of operation, is used to inject a balancing current I into the bridge circuit at one of the nodes. $R1$ is adjusted, with the stimulus at zero, for zero output voltage.

The best calibration method is to apply a precisely known value of stimulus to the transducer, and adjust the amplifier following the transducer for the output proper for that level of stimulus. But that may prove unreasonably difficult in some cases, so an *artificial* calibrator is needed to simulate the stimulus. This function is provided by $R3$ and $S1$ in Figure 13-4. When $S1$ is open, the transducer is able to operate normally, but when $S1$ is closed it *unbalances* the bridge and produces an output voltage E_0 that simulates some standard value of the stimulus. The value of $R3$ is given by

Section 13.9 / BALANCING AND CALIBRATING THE BRIDGE

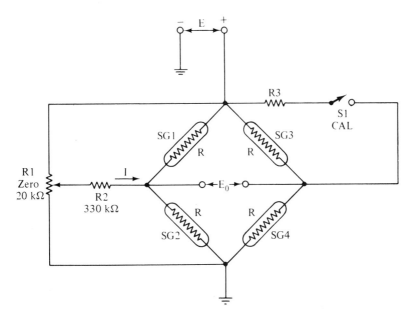

FIGURE 13-4. Circuit for using Wheatstone bridge transducer.

$$R3 = \left[\frac{R}{4Q\psi} - \frac{R}{2}\right] \tag{13-16}$$

where $R3$ = the resistance of R3 in ohms
R = the nominal resistance of the bridge arms in ohms
Q = the calibrated stimulus parameter
ψ = the sensitivity factor in $\mu V/V/Q$ (*Note* the difference in the units of ψ: V instead of μV)

Example 13-6

An arterial blood pressure transducer has a sensitivity of 10 $\mu V/V/$Torr, and a nominal bridge arm resistance of 200 ohms. Find a value for $R3$ in Fig. 13-4 if we want the calibration to simulate an arterial pressure of 200 mm Hg (i.e., 200 Torr).

Solution

$$R3 = \left[\frac{R}{4Q\psi} - \frac{R}{2}\right] \tag{13-16}$$

$$R3 = \frac{200\ \Omega}{4 \times \left(\frac{10^{-5}\ V}{V - T}\right) \times 200\ T} - \frac{200\ \Omega}{2}$$

$$R3 = \frac{200\ \Omega}{(4)(10^{-5})(200)} - 100\ \Omega = 24{,}900\ \Omega$$

13.10 TEMPERATURE TRANSDUCERS

A large number of physical phenomena are temperature dependent, so we find quite a variety of electrical temperature transducers on the market. In this discussion, however, we will discuss only three basic types: *thermistor*, *thermocouple*, and *semiconductor pn junctions*.

13.11 THERMISTORS

Metals and most other conductors are temperature sensitive, and will change electrical resistance with changes in temperature, namely:

$$R_t = R_0[1 + \alpha(T - T_0)] \qquad (13\text{-}17)$$

where R_t = the resistance in ohms at temperature T

R_0 = the resistance in ohms at temperature T_0 (often a standard reference temperature)

T = the temperature of the conductor

T_0 = a previous temperature of the conductor at which R_0 was determined.

α = the *temperature coefficient* of the material, a property of the conductor, in $°C^{-1}$

The temperature coefficients of most metals is positive, as are the coefficients for most semiconductors; e.g., gold has a value of $+0.004/°C$. Ceramic semiconductors used to make *thermistors*, i.e., thermal resistors, can have either negative or positive temperature coefficients depending upon their composition.

The resistance of a thermistor is given by

$$R_t = R_0 e^{\beta[(1/T) - 1/T_0)]}$$

where R_t = the resistance of the thermistor at temperature T

R_0 = the resistance of the thermistor at a reference temperature (usually the ice point, 0°C, or room temperature, 25°C)

e = the base of the natural logarithms
T = the thermistor temperature in degrees Kelvin (°K)
T_0 = the reference temperature in degrees Kelvin (°K)
β = a property of the material used to make the thermistor

(*Note:* β will usually have a value between 1500°K and 7000°K.)

Example 13-7

Calculate the resistance of a thermistor at 100°C if the resistance at 0°C was 18 kΩ. The material of the thermistor has a value of 2200°K. (*Note:* 0°C = 273°K, so 100°C = 373°K.)

Solution

$$R_t = R_0 \exp \beta \left(\frac{1}{T} - \frac{1}{T_0}\right) \quad (13\text{-}18)$$

$$R_t = (1.8 \times 10^4 \, \Omega) \exp \left[(2200°K)\left(\frac{1}{373°K} - \frac{1}{273°K}\right)\right]$$

$$R_t = 1.8 \times 10^4 \, \Omega \, e^{-2.15} = \mathbf{2089 \, \Omega}$$

Equation (13-18) demonstrates that the response of a thermistor is exponential, as shown in Figure 13-5. Note that both curves are nearly linear over a portion of their ranges, but become decidedly nonlinear in the remainder of the region. If a wide measurement range is needed, then a linearization network will be required.

Thermistor transducers will be used in any of the circuits in Figure 13-3. They will also be found using many packaging arrangements. Figure 13-6 shows a bead thermistor used in medical instruments to continuously monitor a patient's rectal temperature.

The equations governing thermistors usually apply if there is little *self-heating* of the thermistor, although there are applications where self-heating is used. But in straight temperature measurements it is to be avoided. To minimize self-heating it is necessary to control the power dissipation of the thermistor.

Also of concern in some applications is the *time constant* of the thermistor. The resistance does not jump immediately to the new value when

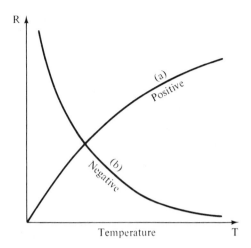

FIGURE 13-5. Thermistor temperature-vs-resistance curves.

the temperature changes, but requires a small amount of time to stabilize at the new resistance value. This is expressed in terms of the time constant of the thermistor in a manner that is reminiscent of capacitors charging in RC circuits.

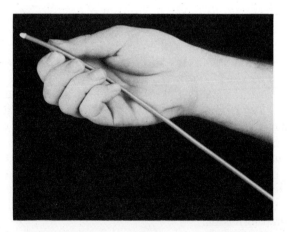

FIGURE 13-6. Thermistor in a medical rectal probe. (*Courtesy of Electronics-for-Medicine.*)

13.12 THERMOCOUPLES

When two dissimilar metals are joined together to form a "vee" [as in Figure 13-7(a)], it is possible to generate an electrical potential merely by heating the junction. This phenomenon, first noted by Seebeck in 1823, is due to different *work functions* for the two metals. Such a junction is called a *thermocouple*. Seebeck EMF generated by the junction is proportional to the junction temperature, and is reasonably linear over wide temperature ranges.

A simple thermocouple is shown in Figure 13-7(b), and uses two junctions. One junction is the measurement junction, and it is used as the thermometry probe. The other junction is a reference, and is kept at a reference temperature such as the ice point (0°C) or room temperature.

Interestingly enough, there is an inverse thermocouple phenomenon, called the *Peltier effect*, in which an electrical potential applied across A-B in Figure 13-7(b) will cause one junction to absorb heat (i.e., get hot) and the other to lose heat (i.e., get cold). Semiconductor thermocouples have been used in small-scale environmental temperature chambers, and it is reported that one company researched the possibility of using Peltier devices to cool submarine equipment. Ordinary air conditioning equipment proves too noisy in submarines desirous of "silent running."

13.13 SEMICONDUCTOR TEMPERATURE TRANSDUCERS

Ordinary pn junction diodes exhibit a strong dependence upon temperature. This effect can be easily demonstrated by using an ohmmeter and an

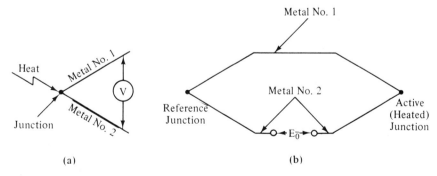

FIGURE 13-7. (a) Thermocouple junction. (b) Two-thermocouple temperature transducer.

ordinary rectifier diode such as the 1N4000-series devices. Connect the ohmmeter so that it forward-biases the diode, and note the resistance at room temperature. Next hold a soldering iron or other heat source close to the diode's body, and watch the electrical resistance change. In a circuit such as Figure 13-8 the current is held constant, so output voltage E_0 will change with temperature-caused changes in diode resistance.

Another solid-state temperature transducer is shown in Figure 13-9. In this version, the temperature sensor device is a pair of diode-connected transistors. In any transistor the base-emitter voltage V_{be} is

$$V_{be} = \frac{kT}{q} \ln \left[\frac{I_c}{I_s}\right] \tag{13-19}$$

where V_{be} = the base-emitter potential in volts (V)
 k = Boltzmann's constant (1.38 × 10^{-23} J/°K)
 T = the temperature in degrees Kelvin (°K)
 Q = the electronic charge (1.6 × 10^{-19} coulomb) ln denotes the natural logarithms
 I_c = the collector current in amperes (A)
 I_s = the reverse saturation current in amperes (A)

Note that the k and q terms in Equation (13-19) are constants, and both currents can be made to be constant. The only variable, then, is *temperature*.

In the circuit of Figure 13-9 we use two transistors connected to provide a differential output voltage ΔV_{be} that is the difference between

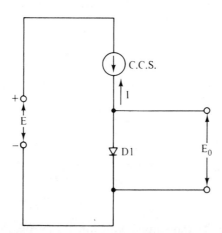

FIGURE 13-8. PN junction diode as a temperature transducer.

Section 13.3 / SEMICONDUCTOR TEMPERATURE TRANSDUCERS

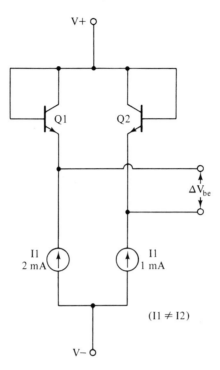

FIGURE 13-9. Two transistors connected as diodes form a temperature transducer if I1 ≠ I2.

$V_{be(Q1)}$ and $V_{be(Q2)}$. Combining the expressions for V_{be} for both transistors yields the expression

$$\Delta V_{be} = \frac{kT}{q} \ln \left[\frac{I1}{I2}\right] \quad (13\text{-}20)$$

Note that, since ln 1 = 0, currents $I1$ and $I2$ *must not be equal*. In general, designers set a ratio of 2:1, i.e., $I1$ = 2 mA and $I2$ = 1 mA. Since currents $I1$ and $I2$ are supplied from constant current sources, the ratio $I1/I2$ is a constant. Also, it is true that the logarithm of a constant is a constant. Therefore, all terms in Equation (13-20) are constants, except temperature T. Equation (13-20), therefore, may be written in the form

$$\Delta V_{be} = KT \quad (13\text{-}21)$$

where $K = (k/q) \ln (I1/I2)$
$K = (1.38 \times 10^{-23})/(1.6 \times 10^{-19}) \ln (2/1)$
$K = 5.98 \times 10^{-5} \text{ V/°K} = 59.8 \ \mu\text{V/°K}$

We may now rewrite Equation (13-21) in the form

$$\Delta V_{be} = 59.8 \ \mu V/°K$$

Example 13-8

Calculate the output voltage from a circuit such as Figure 13-9 if the temperature is 35°C (*Hint:* °K = °C + 273)

Solution

$$\Delta V_{be} = KT \qquad (13\text{-}21)$$

$$\Delta V_{be} = \frac{59.8 \ \mu V}{°K} \times (35 + 273)°K$$

$$\Delta V_{be} = (59.8)(308) \ \mu V = 18{,}418 \ \mu V = \mathbf{0.0184 \ V}$$

In most thermometers using the circuit of Figure 13-9 an amplifier increases the output voltage to a level that is numerically the same as a unit of temperature, so that the temperature may be easily read from a digital voltmeter. The most common scale factor is 10 mV/°K, so for our transducer, the post-amplifier requires a gain of

$$A_v = \frac{10 \ mV/°K}{59.8 \ \mu V \times \frac{1 \ mV}{10^3 \ \mu V}} = 167$$

13.14 INDUCTIVE TRANSDUCERS

Inductance L and inductive reactance X_L are transducible properties, because they can be *varied* by certain mechanical methods.

Figure 13-10(a) shows an example of an inductive Wheatstone bridge. Resistors $R1$ and $R2$ form two fixed arms of the bridge, while coils $L1$ and $L2$ form variable arms. Since inductors are used, the excitation voltage must be ac. In most cases, the ac excitation source will have a frequency between 400 and 5000 Hz, and an rms amplitude of 5 to 10 volts.

The inductors are constructed coaxially, as shown in Figure 13-10(b), with a common core. It is a fundamental property of any inductor that a ferrous core increases its inductance. In the rest condition, i.e., zero-stimulus, the core will be positioned equally inside of both coils.

Section 13.15 / LVDT TRANSFORMERS

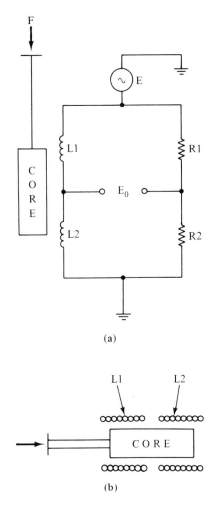

FIGURE 13-10. (a) Inductive Wheatstone bridge transducer. (b) Mechanical form.

If the stimulus moves the core in the direction shown in Figure 13-10(b), the core tends to move out of $L1$ and further into $L2$. This action reduces the inductive reactance of $L1$ and increases that of $L2$, unbalancing the bridge.

13.15 LINEAR VARIABLE DIFFERENTIAL TRANSFORMERS (LVDT)

Another form of inductive transformer is the *linear variable differential transformer* (LVDT) shown in Figure 13-11. The construction of the

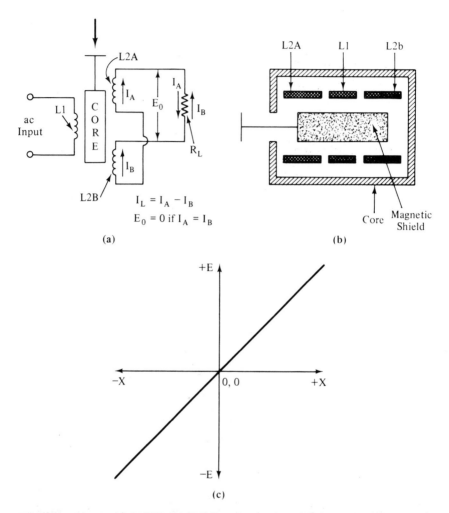

FIGURE 13-11. (a) LVDT. (b) LVDT construction. (c) Output transfer function.

LVDT is similar to that of the inductive bridge, except that it also contains a primary winding.

One advantage of the LVDT over the bridge-type transducer is that it provides higher output voltages for small changes in core position. Several commercial models are available that produce 50 mV/mm to 300 mV/mm. In the latter case, this means that a 1-mm displacement of the core produces a voltage output of 300 mV.

In normal operation, the core is equally inside both secondary coils,

$L2a$ and $L2b$, and an ac carrier is applied to the primary winding. This carrier typically has a frequency between 40 Hz and 20 kHz, and an amplitude in the range 1 V rms and 10 V rms.

Under rest conditions the coupling between the primary and each secondary is equal. The currents flowing in each secondary, then, are equal to each other. Note in Figure 13-11(a) that the secondary windings are connected in series-opposing, so if the secondary winding currents are equal, they will exactly cancel each other in the load. The ac voltage appearing across the load, therefore, is *zero* ($I_a = I_b$).

But when the core is moved so that it is more inside $L2b$ and less inside $L2a$, the coupling between the primary and $L2b$ is greater than the coupling between the primary and $L2a$. Since this fact makes the two secondary currents no longer equal, the cancellation is not complete. The current in the load I_L is no longer zero. The output voltage appearing across load resistor R_L is proportional to the core displacement, as shown in Figure 13-11(c). The *magnitude* of the output voltage is proportional to the *amount* of core displacement, while the *phase* of the output voltage is determined by the *direction* of the displacement.

13.16 POSITION-DISPLACEMENT TRANSDUCERS

A position transducer will create an output signal that is proportional to the position of some object along a given axis. For very small position ranges we could use a strain gage (i.e., Figure 13-12), but note that the range of such transducers is necessarily very small. Most strain gages either are nonlinear for large displacements, or are damaged by large displacements.

The LVDT can be used as a position transducer. Recall that the output polarity indicates the direction of movement from a zero-reference position, and the amplitude indicates the magnitude of the displacement. Although the LVDT will accommodate larger displacements than the strain gage, it is still limited in range.

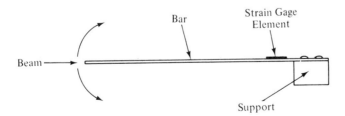

FIGURE 13-12. Beam transducer using a strain gage element.

The most common form of position transducer is the potentiometer. For applications that are not too critical, it is often the case that ordinary linear taper potentiometers are sufficient. Rotary models are used for curvilinear motion, and slide models for rectilinear motion.

We have already seen an example of the potentiometer as position transducers in our discussion (Chapter 10) on servomechanism strip-chart recorders.

In precision applications designers use either regular precision potentiometers or special potentiometers designed specifically as position transducers.

Figure 13-13 shows two possible circuits using potentiometers as position transducers. In Figure 13-13(a) we see a single-quadrant circuit for use where the zero point, i.e., starting reference, is at one end of the scale. The pointer will always be at some point such that $0 \leq x \leq X_m$. The potentiometer is connected so that one end is grounded and the other is connected to a precision, regulated voltage source $V+$. The value of V_x represents X, and will be $0 \leq V_x \leq V+$, such that $V_x = 0$ when $X = 0$, and $V_x = V+$ when $X = X_m$.

A two-quadrant system is shown in Figure 13-13(b), and is similar to the previous circuit except that instead of grounding one end of the potentiometer, it is connected to a precision, regulated *negative*-to-ground power source, $V-$. Figure 13-14 shows the output functions of these two transducers. Figure 13-14(a) represents the circuit of Figure 13-13(a), while Figure 13-14(b) represents the circuit of Figure 13-13(b).

A four-quadrant transducer can be made by placing two circuits such as Figure 13-13(b) at right angles to each other, and arranging linkage so that the output signal varies appropriately.

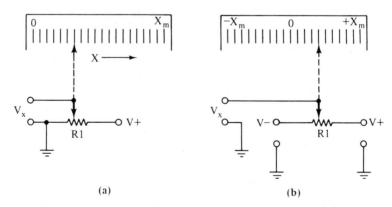

FIGURE 13-13. (a) Position transducer using a potentiometer (one quadrant). (b) Position transducer using a potentiometer (two quadrants).

Section 13.17 / VELOCITY AND ACCELERATION TRANSDUCERS

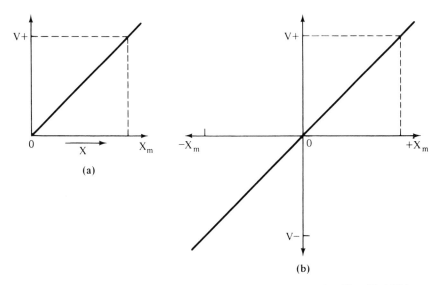

FIGURE 13-14. (a) *V-vs-X* for Fig. 13-13(a). (b) *V-vs-X* for Fig. 13-13(b).

13.17 VELOCITY AND ACCELERATION TRANSDUCERS

Velocity can be defined as displacement per unit of time, and acceleration is the time rate of change of velocity. Since both velocity (v) and acceleration (a) can be related back to position (s), we often find position transducers used to *derive* velocity and acceleration signals. The relationships are

$$v = \frac{ds}{dt} \qquad (13\text{-}22)$$

$$a = \frac{dv}{dt} \qquad (13\text{-}23)$$

$$a = \frac{d^2s}{dt^2} \qquad (13\text{-}24)$$

Velocity and acceleration are the first and second time derivatives of displacement (i.e., change of position), respectively. We may derive electrical signals proportional to v and a by using an operational amplifier differentiator circuit (see Figure 13-15). The output of the transducer is a time-dependent function of position, i.e., displacement. This signal is dif-

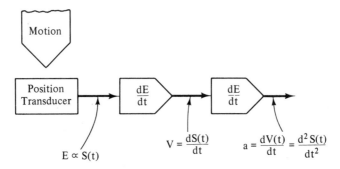

FIGURE 13-15. Example of derived signals using integration or differentiation.

ferentiated by the stages following to produce the velocity and acceleration signals.

13.18 TACHOMETERS

AC and dc generators are also used as velocity transducers. In their basic form they will transduce rotary motion—i.e., produce an angular velocity signal—but with appropriate mechanical linkage will also indicate rectilinear motion.

In the case of a dc generator, the output signal is a dc voltage with a magnitude that is proportional to the angular velocity of the armature shaft.

The ac generator, or *alternator*, maintains a relatively constant output voltage, but its ac *frequency* is proportional to the angular velocity of the armature shaft.

If a dc output is desired, instead of an ac signal, then a circuit similar to Figure 13-16 is used. The ac output of the tachometer is fed to a trigger

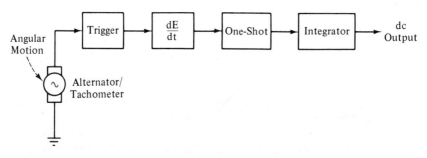

FIGURE 13-16. Alternator tachometer.

Section 13.19 / FORCE AND PRESSURE TRANSDUCERS

circuit (either a comparator or Schmitt trigger) so that squared-off pulses are created. These pulses are then differentiated to produce spike-like pulses to trigger the monostable multivibrator (one-shot). The output of the one-shot is integrated to produce a dc level proportional to the tachometer frequency.

The reason for using the one-shot stage is to produce output pulses that have a *constant amplitude* and *duration*. Only the pulse repetition rate (i.e., number of pulses per unit of time) varies with the input frequency. This fact allows us to integrate the one-shot output to obtain our needed dc signal. If either duration or amplitude varied, then the integrator output would be meaningless. This technique, incidentally, is widespread in electronic instruments, so it should be understood well.

13.19 FORCE AND PRESSURE TRANSDUCERS

Force transducers can be made by using strain gages, or either LVDT or potentiometer displacement transducers. In the case of the displacement transducer (Figure 13-17(a)) it becomes a force transducer by causing a power spring either to compress or stretch. Recall Hooke's law, which tells us that the force required to compress or stretch a spring is proportional to a *constant* and the *displacement* caused by the compression or tension force applied to the spring. So by using a displacement transducer and a calibrated spring, we are able to measure force.

Strain gages connected to flexible metal bars (refer back to Figure 13-12) are also used to measure force, because it requires a certain amount of force to deflect the bar any given amount. There are several transducers on the market that use this technique, and they are advertised as "force-displacement" transducers. Such transducers form the basis of the digital bathroom scales now on the market.

Do not be surprised to see such transducers, especially the smaller types, calibrated in *grams*. We all know that the gram is a unit of *mass*, not force, so what this usage refers to is the gravitational force on one gram at the earth's surface, roughly 980 dynes. A one-g weight suspended from the end of the bar in Figure 13-12 will represent a force of 980 dynes.

A side view of a cantilever force transducer is shown in Figure 13-17(b). In this device a flexible strip is supported by mounts at either end, and a piezoresistive strain gage is mounted to the under side of the strip. Flexing the strip unbalances the gage's Wheatstone bridge, producing an output voltage.

A related device uses a cup- or barrel-shaped support, and a circular diaphragm instead of the strip. Such a device will measure force or pressure, i.e., *force per unit of area*.

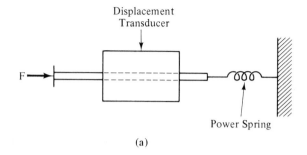

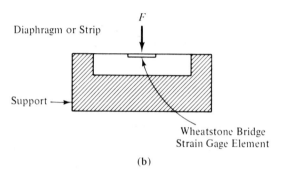

FIGURE 13-17. (a) Force from a displacement transducer. (b) Force/pressure transducer.

13.20 FLUID PRESSURE TRANSDUCERS

Fluid pressures are measured in a variety of ways, but the most common involve a transducer such as those shown in Figures 13-18 and 13-19.

In the example of Figure 13-18(a) a strain gage or LVDT is mounted inside a housing that has a bellows or aneroid assembly exposed to the fluid. More force is applied to the LVDT or gage assembly as the bellows compresses. The compression of the bellows is proportional to the fluid pressure.

An example of the *Bourdon tube* pressure transducer is shown in Figure 13-18(b). Such a tube is hollow and curved, but flexible. When a pressure is applied through the inlet port, the tube tends to straighten out. If the end tip is connected to a position/displacement transducer, then the transducer output will be proportional to the applied pressure.

Figure 13-19(a) shows another popular form of fluid transducer. In this version, a diaphragm is mounted on a cylindrical support similar to Figure 13-18. In some cases, a bonded strain gage is attached to the under

Section 13.20 / FLUID PRESSURE TRANSDUCERS

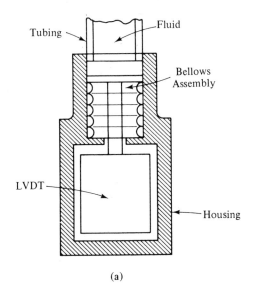

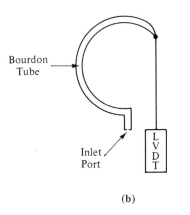

FIGURE 13-18. (a) Fluid pressure transducer. (b) Bourdon tube fluid pressure transducer.

side of the diaphragm, or flexible supports to an unbonded type are used. In the example shown, the diaphragm is connected to the core drive bar of an inductive transducer or LVDT.

Figure 13-19(b) shows the Hewlett-Packard type 1280 transducer used in medical electronics to measure human blood pressure. In this device, the hollow fluid-filled dome is fitted with *Luer-lock* fittings, standards in medical apparatus.

The fluid transducers shown so far will measure gage pressure, i.e.,

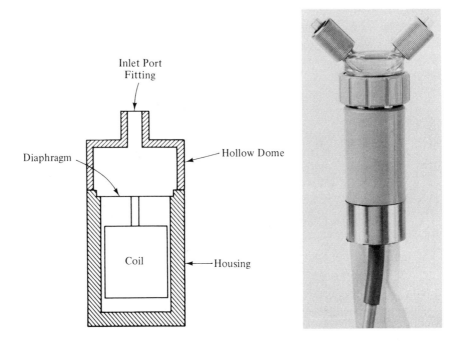

FIGURE 13-19. (a) Dome-type fluid pressure transducer. (b) Commercial dome-type transducer. (*Courtesy of Hewlett-Packard.*)

pressure above atmospheric pressure, because one side of the diaphragm is open to air. A *differential* pressure transducer will measure the difference between pressures applied to the two sides of the diaphragm. Such devices will have two ports marked such as "$P1$" and "$P2$," or something similar.

13.21 LIGHT TRANSDUCERS

There are several different phenomena that can be used for measuring light, and they create different types of transducer. For this chapter, we will limit the discussion to *photoresistors, photovoltaic cells, photodiodes,* and *phototransistors.*

A photoresistor can be made because certain semiconductor elements show a marked decrease in electrical resistance when exposed to light. Most materials do not change linearly with increased light intensity, but certain combinations such as cadmium sulphide (CdS) and cadmium

FIGURE 13-20. (a) Symbol for photoresistor cell. (b) Actual photoresistor cell.

selenide (CdSe) are effective. These cells operate over a spectrum from "near-infrared" through most of the visible light range, and can be made to operate at light levels of 10^{-3} to 10^{+3} footcandles (i.e., 10^{-3} to 70 mW/cm²). Figure 13-20(a) shows the photoresistor circuit symbol, while Figure 13-20(b) shows an example of a photoresistor.

A photovoltaic cell, or "solar cell," as it is sometimes called, will produce an electrical current when connected to a load. Both silicon (Si) and selenium (Se) types are known. The Si type covers the visible and near-infrared spectrum, at intensities between 10^{-3} and 10^{+3} mW/cm². The selenium cell, on the other hand, operates at intensities of 10^{-1} to 10^2 mW/cm², but accepts a spectrum of near-infrared to the ultraviolet.

Semiconductor pn junctions under sufficient illumination will respond to light. Interestingly enough, they tend to be photoconductive when heavily reverse-biased, and photovoltaic when forward-biased. These phenomena have led to a whole family of photodiodes and phototransistors.

13.22 CAPACITIVE TRANSDUCERS

A parallel plate capacitor can be made by positioning two conductive planes parallel to each other. The capacitance is given by

$$C = \frac{kKA}{d} \tag{13-25}$$

where C = the capacitance in *farads* (F), or a subunit (μF, pF, etc.)
k = a units constant
K = the dielectric constant of the material used in the space between the plates (K for air is 1)
A = the area of the plates "shading" each other
d = the distance between the plates

Figure 13-21 shows several forms of capacitance transducer. In Figure 13-21(a) we see a rotary plate capacitor that is not unlike the variable capacitors used to tune radio transmitters and receivers. The capacitance of this unit is proportional to the amount of area on the fixed plate that is covered, i.e., "shaded," by the moving plate. This type of transducer will give signals proportional to curvilinear displacement, or angular velocity.

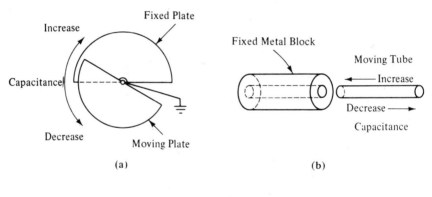

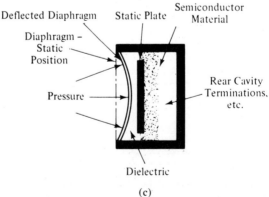

FIGURE 13-21. Capacitance transducers. (*Source:* Harry Thomas, *Handbook of Biomedical Instrumentation and Measurement,* Fig. 1-6, p. 13, Reston Publishing Co.)

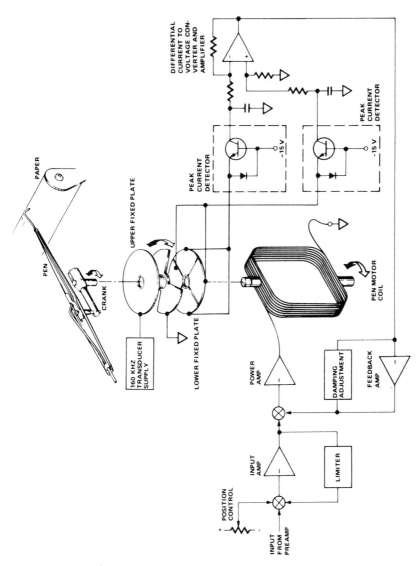

FIGURE 13-22. Capacitance transducers used as a position/velocity sensor. *(Courtesy of Hewlett-Packard.)*

A rectilinear capacitance transducer is shown in Figure 13-21(b), and it consists of a fixed cylinder and a moving cylinder. These pieces are configured so that the moving piece fits inside of the fixed piece, but is insulated from it.

Two types of capacitive transducer discussed so far vary capacitance by changing the shaded area of two conductive surfaces. Figure 13-21(c), on the other hand, shows a transducer that varies the *spacing* between surfaces, i.e., the d term in Equation (13-25). In this device, the metal surfaces are a fixed plate and a thin diaphragm. The dielectric is either air or a vacuum. Such devices are often used as capacitance microphones.

Capacitance transducers can be used in several ways. One method is to use the varying capacitance to frequency-modulate an RF oscillator. This is the method employed with capacitance microphones [those built like Figure 13-21(c), not the electrotet type]. Another method is to use the capacitance transducer in an ac bridge circuit.

Figure 13-22 shows an application in which a differential capacitance transducer is used to provide a position signal in a PMMC galvanometer chart recorder (Chapter 10). This type of transducer is superior to others in this application, because it can be constructed with very low mass, so it will not significantly dampen the pen motion. The position signal is used in a feedback control circuit that gives the pen controlled damping. In one model, the differentiated position signal warns the instrument that the pen will slam into the mechanical limit stop, and puts the brakes on to prevent pen damage.

13.23 SUMMARY

1. A *transducer* is a device that converts a physical stimulus such as displacement, force, temperature, etc., to an electrical voltage or current proportional to the magnitude of the stimulus.
2. A *strain gage* depends upon the *piezoresistance*—i.e., resistance change due to mechanical deformation—phenomenon.
3. There are two basic forms of inductive transducer: X_L bridge and LVDT.
4. Capacitive transducers depend upon varying the space between, or the shaded area of, two conductive surfaces.
5. Several types of temperature transducer are common: thermistor, thermocouple, and semiconductor junctions.

13.24 RECAPITULATION

Now go back and try answering the questions at the beginning of the chapter. When you are finished, answer the questions and work the

Section 13.25 / QUESTIONS

problems given below. Place a mark beside each problem or question that you cannot answer, and then go back to the text and reread appropriate sections.

13.25 QUESTIONS

1. Define in your own words "transducer."
2. List several electrical parameters that can be used to transduce physical parameters.
3. Many transducers are often used in _____ bridge circuits.
4. Define "piezoresistivity" in your own words.
5. *Tension* forces in a cylindrical conductor _____ the electrical resistance.
6. *Compression* forces in a cylindrical conductor _____ the electrical resistance.
7. Changes in electrical resistance due to deformation are a phenomenon called _____.
8. Define in your own words "gage factor."
9. Write the two equations for gage factor.
10. Discuss the differences between *bonded* and *unbonded* strain gages.
11. Write the equation for E_0 in a Wheatstone bridge transducer that uses *one* active element.
12. What are the *units* for the sensitivity of the fluid pressure transducer?
13. Draw a circuit showing a *balance* control in a Wheatstone bridge strain gage.
14. List three types of temperature transducer.
15. Define *Seebeck effect* and *Peltier effect* in your own words. How are these effects related?
16. The temperature coefficients for most elemental metals are _____.
17. Specially "doped" semiconductor materials can have both _____ and _____ temperature coefficients.
18. A thermistor used as an electronic thermometer must be operated to minimize _____.
19. A thermocouple is formed of two _____ metals, and works because the metals have different _____.
20. List three types of semiconductor temperature transducer.
21. List two types of inductive transducer. Draw appropriate circuit diagrams.
22. The secondary windings of an LVDT are connected in _____.

23. List two types of position/displacement transducers.
24. Draw circuits of one- and two-quadrant position transducers.
25. How can a displacement transducer derive velocity and acceleration signals?
26. What types of device are used to measure angular velocity?
27. List three types of phototransducer.
28. List two types of photoconductive semiconductor element.

13.26 PROBLEMS

1. A constantan cylinder with a diameter of 0.05 mm is 14.6 cm long. Find its electrical resistance.
2. Predict the *gage factor* if a 12-mm-long strain gage element changes length 0.7 mm, and its resistance changes from 78 to 82 ohms. What type of force is applied, *tension* or *compression?*
3. What is the *gage factor* of a cylindrical wire 0.05 mm in diameter and 18 mm long if the diameter changes 0.03 mm and the length changes 2.9 mm when a tension force is applied?
4. A Wheatstone bridge strain gage is excited by a 6-volt dc source. When it is stimulated by an outside parameter each element changes resistance by a factor of $0.042R$, and all four arms have equal resistance under zero-stimulus conditions. Calculate the output voltage E_0 for (a) one, (b) two, and (c) three active elements.
5. A 5-gram mass hangs from the end of a bar-type strain gage (i.e., Figure 13-12). Express this mass as a force in *dynes*.
6. A fluid pressure transducer has a sensitivity of 50 μV/V/cm Hg and is excited by a +7.5-volt dc source. Find the output voltage if a 150-mm Hg pressure is applied.
7. The transducer in Problem 6 is a Wheatstone bridge strain gage, with four active elements that each have a zero-pressure resistance of 500 ohms. Calculate the value of a calibration resistance (to shunt one arm) that will give an artificial pressure of 120 mm Hg.
8. A thermistor is known to have a resistance of 100 kΩ at 25°C, and a temperature coefficient of $0.02°C^{-1}$. Calculate the resistance at (a) 30°C and (b) 56°C.
9. Find the resistance of a thermistor at 100°C if the ice-point resistance is 500 kΩ, and the value of β is 2900°K.
10. Find the base-emitter voltage of a transistor at 33°C if the collector current is 10 mA, and the reverse saturation current is 10^{-13} ampere.
11. Find the slope factor of a dual-transistor temperature transducer such as Figure 13-9 if $I1 = 10$ mA and $I2 = 3$ mA.
12. What amplification following the transducer in Problem 11 results in a scale factor of 10 mV/°K?

13.27 DESIGN PROJECTS

The design projects in this section are practical examples of how the chapter material might be applied. There may be more than one way to solve each problem, and you must convince the instructor that your method works.

13-1. Design an analog sub-system for a microcomputer that will measure temperature in degrees *centigrade*. Do not provide for A/D conversion unless the instructor has already covered Chapters 17-20. The user does not want to buy new transducers, and has been using a dual-transistor PN junction. Provide a scale factor of 10 mV/°K, and a readout that is zero volts at 0°C.

13-2. A fluid pressure transducer has a range of -250 to $+1500$ cm H_2O. Design an analog subsystem that will output a voltage to the computer's analog-to-digital converter of 10 mV/Torr. (*hint:* 1 T = 1 mm H_g = 13.56 cm H_2O). Do not include an analog-to-digital converter unless the instructor has already covered chapters 17-20.

13-3. A force-displacement transducer is a resistive Wheatstone bridge with a sensitivity of 5 uV/V/gm,[1] and has a range of 0 to 250 grams.[1] Design an analog subsystem that uses DC excitation and will provide a range of output voltage of 0 to $+2.5$ volts for the stated force range.

13-4. A boiler used to heat a hospital is being retrofitted with a microcomputer controller. The purpose of the controller is to turn on the oil burner flame either (a) when the temperature from a central thermostat drops below 40 °F, *or*, (b) once every 20 minutes. A flame spotter is required in order to insure that the flame actually comes on so that fuel is not spilled dangerously into the fire box following nonignition.
    1. Design an analog sub-system that will solve the interface problems for this project. Use a photoresistive cell for the flame spotter. No analog/digital converter is needed for this project.
    2. Write either a Z80 or 6502 program that will recognize the various states of the controller and take action as follows:
        a. Print "BURNER ON—THERM" for a thermostat initiated stimulus.
        b. Print "BURNER ON—TIME" for a timer initiated stimulus.

---

[1] *Note: Grams* are normally considered units of *mass*. In the life sciences, however, they are often used as units of *force* because the transducer can then easily be calibrated by the use of the precision mass elements that are normally used to calibrate balance scales. The actual unit of force in this case is the *dyne*, in which one gram-force is the gravitational attraction for one gram-mass, of about 980 dynes.

c. Print "FLAME-OUT" 10 times if the flame detector indicates an alarm condition.
d. Turn off the power-on relay to the furnace if a flame-out alarm is received.

13.28 REFERENCES

1. W.D. Cooper, *Electronic Instrumentation and Measurement Techniques*. Englewood Cliffs, N.J.: Prentice-Hall, Inc., 1970.
2. Frank J. Oliver, *Practical Instrumentation Transducers*. New York: Hayden Book Co., 1971.
3. Peter Strong, *Biophysical Measurements*. Beaverton, Ore.: Tektronix, Inc., Measurement Concepts Series, 1973.
4. Sol Prensky, *Electronic Instrumentation,* 2nd ed. Englewood Cliffs, N.J.: Prentice-Hall, Inc., 1971.
5. Joseph Tusinski, "Strain Gages Come of Age," *Electronics World* (March, 1969), pp. 35–37.
6. *Reference Data for Radio Engineers*. Indianapolis, Ind.: Howard W. Sams & Co., Inc., 1968.
7. D.H. Sheingold, ed., *Nonlinear Circuits Handbook*. Norwood, Mass.: Analog Devices, Inc., 1974.
8. Richard S.C. Cobbold, *Transducers for Biomedical Measurements*. New York: Wiley-Interscience, 1974.
9. Harry E. Thomas, *Handbook of Biomedical Instrumentation and Measurements*. Reston, Va.: Reston Publishing Co., 1974.

14
Sample & Hold Circuits

14.1 OBJECTIVES

1. To learn the basic principles of Sample & Hold circuits.
2. To understand design problems in the S&H circuit.
3. To learn the use of S&H circuits in computer-based instruments.
4. To examine some commercial IC S&H circuits.

14.2 SELF-EVALUATION QUESTIONS

Before studying the material in this chapter, try answering the questions given below. These questions test your prior knowledge of the subject matter. If you cannot answer a particular question, then place a check mark beside it and look for the answer as you read the text.

1. What type of transistor switches are typically used in Sample & Hold circuits.
2. What property is most important in the operational amplifier used as an S&H output buffer?
3. Define S&H *settling time*.
4. What principal defect occurs in some S&H circuits if the hold time is too long?
5. What are typical uses for S&H circuits in microcomputer data acquisition?

14.3 INTRODUCTION

The *Sample & Hold* circuit is designed to take a brief look at an analog signal and hold its value indefinitely. Although certain practical design problems limit the term *indefinitely* to a very short duration, the S&H can be used to hold a value at or close to its original value for a period of time after the signal itself has changed value. There seem to be two major uses of the S&H circuit in microcomputer data acquisition systems. First, we find it used at the input of analog-to-digital converters where it is used to keep the analog signal constant during the data conversion process. Some A/D converters are particularly prone to substantial errors if the signal being converted changes. Second, we sometimes see multiple S&H circuits used when a single A/D converter handles many channels. This arrangement allows us to take the digital value of several possibly related parameters. If we took the samples at sequential times, instead of identical times, then the correlation between related values in the different channels is reduced—possibly destroyed.

The use of S&H circuits in non-digital-computer applications is also wide. We sometimes see S&H circuits used to cancel the drift in DC amplifiers by sampling first the signal in one S&H, and then the drift component in a second S&H. This latter job is done with the amplifier input shorted, and is performed during the *hold* period of the signal S&H. We can then subtract out the drift component from the signal by using an ordinary differential operational amplifier.

14.4 ANALOG SWITCHES

The heart of the sample & hold circuit is an analog electronic switch (see Figure 14-1). A typical circuit for a simple analog switch is shown in Figure 14-1(a), while several possible circuit symbols are shown in Figure 14-1(b). In this example, the switching element is a single junction field effect transistor (JFET), while in others (perhaps most) it is either a single Metal Oxide Semiconductor Field Effect Transistor (MOSFET) or a complementary pair of N-channel and P-channel MOSFETs. In any case, the idea is that the transistor is connected in series with the signal path. When the transistor is biased fully off, then no signal will flow. But when the transistor is biased fully on, then the signal easily passes from input to output sides. Typical off/on resistances may be 1,000,000:1.

The usual field effect transistor analog switch contains a switch driver circuit (such as Q2) that will allow a TTL voltage signal to turn the switch on and off. In the example shown in Figure 14-1(a), the driver transistor is used to control the *V*-bias to the gate of the JFET. When the base (control input) of Q2 is LOW, then the base is at a lower potential than the emitter, so the transistor will conduct. The gate of the JFET sees a high negative bias that shuts the switch off. Similarly, when the control line is HIGH, then the transistor is reverse biased and the switch is turned on.

Section 14.5 / SIMPLE S&H CIRCUIT

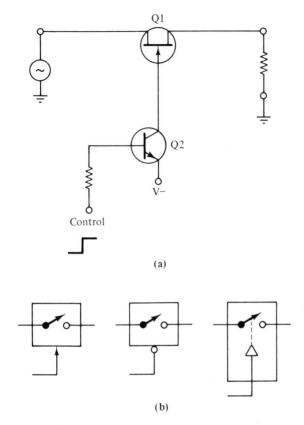

FIGURE 14-1. (a) JFET electronic switch. (b) Electronic switch symbols.

14.5 SIMPLE S&H CIRCUIT

Although there are a number of variations on the basic circuit, we see all of the necessary elements of the sample & hold circuit in Figure 14-2. The main components are an *output buffer amplifier, sample capacitor, sample switch,* and the input circuitry (which may or may not include an *input buffer amplifier*).

The analog sample switch ($S1$) closes during the sample period, allowing the input signal to charge the sample capacitor (C_H). The sample switch then opens up again, leaving the capacitor charged to the value of the analog input signal. The capacitor selected for C_H must be a very low leakage type, or else the voltage will drop off ("droop") due to leakage current across the capacitor dielectric. Typical types used for this are silver mica (not preferred), polycarbonate, polystyrene, or glass (consid-

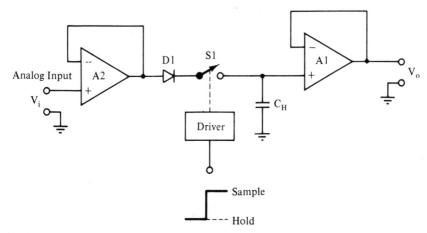

FIGURE 14-2. Sample and hold.

ered obsolete for most uses). It is essential that the capacitor have a very low leakage dielectric.

The buffer amplifier is connected in a unity gain configuration. This operational amplifier circuit has a gain that is slightly less than unity (i.e., 0.99999). The open-loop gain (i.e., the gain without feedback) of the operational amplifier will determine how close to unity the gain will be; the higher A_{vol}, the closer to unity in the unity-gain noninverting circuit shown.

But it is not the A_{vol} that is the most important property of the operational amplifier used for $A1$. After all, reasonably high gain (over 50,000) is available on even the cheapest operational amplifier devices. The most important property of the operational amplifier used as the output buffer is a *high input impedance*. The input impedance of the buffer amplifier represents a load across the sample capacitor that will tend to discharge the capacitor. Also, the input bias currents will tend to affect the charge on C_H. As a result, we must select an operational amplifier with an extremely high input impedance. Superbeta, BiMOS or BiFET input operational amplifiers are a must. The RCA CA3140 device is a good selection because it has an input impedance of 1.5 teraohms (1.5×10^{12} ohms).

The isolation diode is used to keep the charge on the hold capacitor from leaking back through the input load, or the output terminal of the input operational amplifier ($A2$), if used. This diode must be selected to have a low level of reverse leakage current, or excessive droop will occur. It should also have a low value of forward voltage drop. This latter re-

Section 14.5 / SIMPLE S&H CIRCUIT

quirement, however, can be easily compensated for by using a gain circuit for either $A1$ or $A2$. In fact, there are numerous S&H applications where there will be a gain requirement for scaling purposes. In most cases, the input amplifier will have some gain in order to bring the input signal up to a level that will produce an in-range output signal. Besides, it is true that offset bias current errors in the charge of capacitor C_H are lessened in severity if a higher level signal is applied. These errors tend to have a constant value, so will be a smaller percentage of the total if the input signal potential is higher.

Figure 14-3 shows the operation of a typical S&H circuit for a varying input signal, V_{in}. The dotted line represents the input signal, while the solid line represents the output signal V_o. For this example, the sample line must be LOW to sample, and HIGH to hold. Initially, the line is HIGH so the output voltage will remain at the last valid level. In other words, the value is held. But, at time t_1 the S&H line drops LOW, so the input signal is sampled. Time $(t_2 - t_1)$ is the acquisition time. This period must be short enough that the input signal will not slew very far during the sample period; otherwise the data will contain an error. At time t_2 the S&H line snaps HIGH again, which causes the output voltage to remain at the last valid level. This cycle of sample and hold is repeated for different values of signal at time t_3.

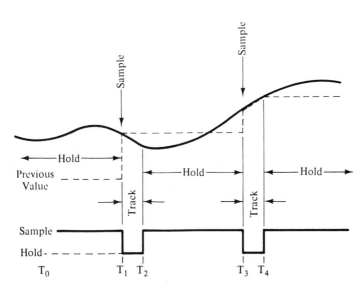

FIGURE 14-3. Sample & hold waveform.

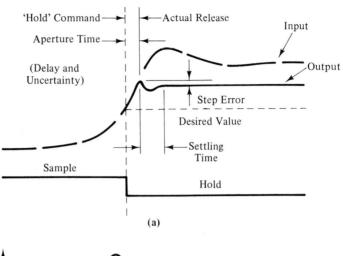
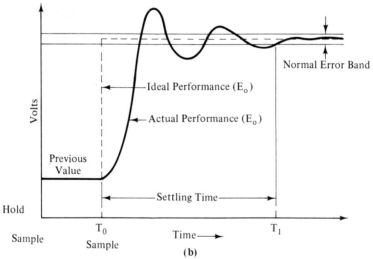
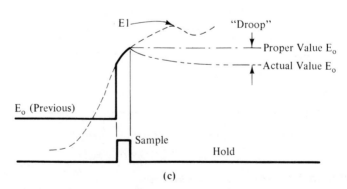

FIGURE 14-4(a). S&H errors. **(b).** Settling time. **(c).** Droop.

14.6 SAMPLE & HOLD ERRORS

While our blackboard circuits are always ideal and error free, real practical circuits often contain substantial errors. Although the individual error in any one circuit is small, the cumulative error, i.e., the summation of all system errors, will be quite large in some cases. As a result, it is necessary for us to understand the errors peculiar to each circuit and their effect on the overall efficacy of the circuit or system that we are designing.

In sample & hold circuits there are at least eight basic forms of errors: these are *aperture time, switching transients, settling time, S-to-H offset voltage, voltage droop, feedthrough, dielectric absorption,* and *gain*.

We can define *aperture time* as the time period between the issuance of the command to sample and the actual closing of the switch (Figure 14-4(a)). We can attribute aperture time to two different classes of problems, switch actuation time and jitter. The switch actuation time is usually constant for any given type of switch, but this time can be substantial when dealing with very rapid S&H circuits. The primary cause of this problem is the charging of the capacitances in the base or gate structures of the switch driver circuits. All such circuits require some charging time. The problem of jitter is similar to the contact bounce problem of mechanical switches.

Aperture time becomes important because it limits the resolution of the circuit for any given input signal slew rate. Consider a hypothetical situation in which we want to resolve the signal to 0.1 percent, and the full-scale input signal is 10 volts. Also assume that the input signal slew rate is 2.5 $V/\mu S$. How do we find the minimum aperture time required for this signal? The equation is:

$$T_a = \frac{V_{fs}r}{S_r} \qquad (14\text{-}1)$$

where:

T_a is the aperture time in seconds (s)

V_{fs} is the fullscale input signal in volts

r is the resolution expressed as a decimal percent (i.e., 0.1 percent is 0.001)

S_r is the slew rate of the input signal in volts per microsecond $(V/\mu S)$

For the problem stated in the text above:

$$T_a = \frac{V_{fs}r}{S_r}$$

$$T_a = \frac{(10.0\ V)(0.001)}{\left[\frac{2.5\ V}{\mu S} \times \frac{10^6\ \mu S}{S}\right]}$$

$$T_a = \frac{0.01}{2.5 \times 10^6/S} = 4 \times 10^{-9}S = \mathbf{4\ nS}$$

For the situation described above, then, we would require a very fast S&H circuit that was capable of an aperture time of less than 4 nanoseconds.

The second type of error is *switching transients*. These occur at the instant of switching, and are of critical importance when the input signal has a high slew rate.

Settling time (Figure 14-4b) is the time required for the Sample & Hold circuit output voltage to permanently settle within the nominal error band specified for the device. This error is usually specified as a given percentage of the fullscale input signal voltage.

Also shown in Figure 14-4a is the *Sample-to-Hold offset error*. This error is due to the charge stored in the capacitances of the FET switch used to take the sample. At the instant that the HOLD mode is instituted, the charge in these capacitances is dumped into the circuit and is stored in the capacitor.

Voltage droop (Figure 14-4c) occurs during the HOLD period, and is due to a slight discharge of the hold capacitor. There will always be a leakage resistance across the capacitor, and a shunt resistance that is due to the input impedance of the output amplifier. The degree of droop is given by Equation (14-2):

$$\frac{dV_o}{dt} = \frac{I}{C_h} \tag{14-2}$$

where:

V_o is the output voltage in volts (V)

I is the capacitor leakage current from all resistances, both internal and external to the capacitor, in amperes (A)

C_h is the capacitance of the S&H capacitor in farads

Output voltage droop is something we must live with, even though

Section 14.6 / SAMPLE & HOLD ERRORS

there are design steps that will mitigate the degree of droop in any given application. We can, for example, be careful to select low leakage capacitors (e.g., polycarbonate and polyethylene), diodes with extremely low leakage currents, and output amplifiers that have extremely high input impedance specs.

Feedthrough is defined as the leakage of signal through the sampling switch. There are two factors to consider here, series resistance in the off state and the capacitance of the open switch (i.e., when in the off state). The presence of any substantial leakage resistance will permit the DC offset component of the input signal to affect the charge on the sampling capacitor. Similarly, the switch capacitance will permit alteration of the charge on the capacitor by AC signals of sufficient frequency.

Dielectric absorption is a property of capacitors that is often overlooked by capacitor users, except in cases such as Sample & Hold circuits and where the capacitor is used at higher voltages (i.e., in medical defibrillators). The electrostatic field used to store energy in the dielectric of a capacitor causes the orbits of the electrons in the atoms of the dielectric to deform. This, in fact, is the manner in which the electrical energy is stored in the capacitor. When the capacitor is discharged the orbits return to their normal shapes; the energy lost in the dielectric by this process becomes the electrical current that flows in the external circuit. But not all electron orbits return to their correct shape at the same time. After the capacitor has discharged there will be a residual charge left in the capacitor. You can see this effect when working with high voltage, oil-filled capacitors that have a high capacitance value (e.g., over 10 uF). Charge the capacitor to several thousand volts (medical defibrillators generally use 16 uF capacitors charged to 7000 volts). After a few seconds attempt to discharge the capacitor—note the spark. Now, take a voltmeter and measure the voltage across the terminals of the supposedly discharged capacitor; you may find over 100 volts remaining. In the small capacitors used in instrumentation S&H circuits this voltage may be smaller than in the more spectacular example above, but it is nonetheless significant to the accuracy of the S&H output signal. This dielectric absorption voltage will be added to the output voltage of the next sample. The dielectric absorption factor is specified for capacitors, and will be from 0.001 percent to 10 percent. Try and select as low a factor as possible; polystyrene is a good selection for most applications, as is *Teflon.*®

Gain error is the error caused by problems with the gains of the operational amplifiers used in the S&H. Most operational amplifiers used in the noninverting unity gain follower configuration have negligible error. But when the amplifier is used in a gain circuit, then the tolerances of the resistors become important and gain errors pop up.

14.7 SUMMARY

1. A sample & hold circuit is designed to take a look at an input signal and hold its value indefinitely.
2. The principle components of the S&H circuit are operational amplifiers, an electronic switch, and a low leakage capacitor.
3. The eight basic errors in S&H circuits are: aperture time, switching transients, settling time, S-to-H offset voltage, voltage droop, feedthrough, dielectric absorption, and gain.

15
Analog Reference Circuits

15.1 OBJECTIVES

1. To learn the requirements placed on dc reference sources in data conversion.
2. To design zener diode references.
3. To learn the various types of IC reference devices available.
4. To learn the limitation of a data converter, viz, the reference source.

15.2 SELF-EVALUATION QUESTIONS

Before studying the material in this chapter, try answering the questions given below. These questions test your knowledge of the subject matter. If you cannot answer a particular question, then place a check mark beside it and look for the answer as you read the text.

1. What are three conditions found in zener diode regulator circuits?
2. Write the equation for series resistance for Zener condition I.
3. Draw the *I-vs-E* curve for a typical 5.6 volt zener diode.
4. Draw the circuit for an operational amplifier reference source.

15.3 INTRODUCTION

The data converters used in microprocessor-based instrumentation are devices that produce an output that is, in part, proportional to a reference source. The simple Digital-to-Analog converter (DAC), for example, produces an output voltage (or current) that is proportional to both the applied binary word *and* an analog reference voltage (or current). The pre-

209

cision and accuracy of the data converter is very often determined by constraints placed on the reference source. Although the bit-length of the converter is normally thought to be the principle limiting factor that determines the resolution of the data, there is also a severe constraint placed on the reference source. For example, if the resolution of a DAC is eight bits, then, we would ordinarily expect that the 1-LSB resolution of the device would be $½^N$, where N is the bit-length. In the case of an eight-bit DAC, then, we would expect the resolution to be the step voltage that is caused by 1-LSB, or $½^8$ which is $1/256$ or 0.0039. Suppose that the full-scale output voltage of the DAC is 2.55 volts. The 1-LSB voltage in that case would be 9.9 millivolts. The reference source should be accurate to 0.99 mV in this case. If it is not, then the 1-LSB resolution of the data converter is in doubt, and one cannot believe the results in that range. It is quite possible to render a data converter no better than another data converter with two or three less bits of word length, just by failing to pay sufficient attention to the matter of the reference source. Consider a typical 10.000 volt output eight-bit DAC. The 1-LSB voltage is 39 millivolts. If the dc reference supply is not adjusted with some precision, or if it drifts off to another voltage with changes in ambient temperature, then the actual 1-LSB voltage will be something different. In one case, the actual voltage was 10.24 volts, instead of 10.000 volts. This eight-bit DAC was incapable of resolving an output potential to more than five bits.

Various types of reference voltage source are used. The simplest, and least desirable, is the simple zener diode source. Other sources, based on either zener diodes or precision versions of the zener such as the band-gap diode, will prove more useful in the development of microprocessor-based instruments. In the sections to follow, we will discuss some of the more popular circuits and IC devices used to make microprocessor data converter references.

15.4 ZENER DIODES

The zener (*pronounced* "zen-ner") diode is a special class of avalanche diode. Figure 15-1(a) shows the typical zener diode regulator circuit, while Figure 15-1(b) shows the response curve typical for a zener diode.

When a zener diode is forward-biased, it behaves very much like any other pn junction diode. Once the forward bias potential exceeds 0.6 to 0.7 volt (its junction potential), the I-vs-E curve becomes essentially linear. But in the reverse-bias condition, the behavior of the zener diode is markedly different from ordinary diodes. The reverse-bias current flow is limited to the normal leakage current, essentially zero or some extremely low value, until the reverse bias potential reaches a critical point called

Section 15.4 / ZENER DIODES

the *zener voltage,* usually symbolized by V_Z. At that point the diode avalanches and a considerable reverse current will flow.

The circuit of Figure 15-1(a) shows a zener diode used as a shunt voltage regulator; i.e., the zener is in parallel with the load. Resistor $R1$ serves to limit the current ($I1$) to a safe value, while $R2$ represents the load resistance.

Voltage regulation occurs if current $I2$ is considerably greater than $I1$; a common rule of thumb is to set $I2$ at approximately 10 times greater than $I3$. If this ratio is maintained, then fluctuations in $I3$ due to changing load conditions become an insignificant percentage of $I1$; therefore, V_Z remains essentially constant.

There are three different circumstances in which the zener circuit of

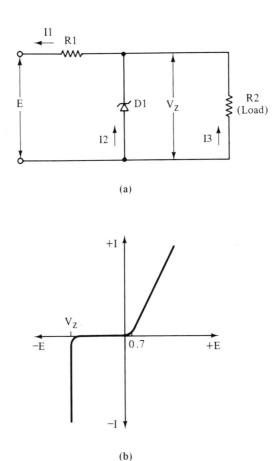

FIGURE 15-1. (a) Zener diode regulator. (b) *I-vs-E* curve for zener diode.

*Condition I**

$$R_s = \frac{E_{(\min)} - V_z}{1.1\,I3}$$

$$P_{D1} = \frac{E_{(\max)} - V_z}{R_s} - I3 \times V_z$$

$$P_{R_s} = P_{D1} + (I3 \times V_z)$$

Condition II

$$R_s = \frac{E - V_z}{1.1\,I3_{(\max)}}$$

Condition III

$$R_s = \frac{E_{(\min)} - V_z}{1.1\,I3_{(\max)}}$$

FIGURE 15-1(c). Design equations for zener diode regulators.

Figure 15-1(a) might be used, and all three have slightly different design equations:

  I. Variable supply voltage (E), constant load current ($I3$)
  II. Constant supply voltage (E), variable load current ($I3$)
  III. Variable supply voltage (E), variable load current ($I3$)

The design equations for these conditions are summarized in Figure 15-1(c). Examples using these equations will be included in the problem set at the end of the chapter.

15.5 PRECISION OPERATIONAL AMPLIFIER DC REFERENCE SUPPLIES

Most dc power supplies produce a specified *nominal* output voltage that is not very precise. A "12-volt dc" supply, for example, might produce a stable output potential somewhere between 11.5 and 12.5 volts, and still be in tolerance. In many applications, however, a more precise voltage is required, such as the reference voltage used to calibrate meters, etc.

The zener diode voltage tends to wander a little with temperature, so in a circuit such as Figure 15-2 the zener diode is placed in a controlled

---

* Use the same power dissipation equations for all three conditions. Subscripts (max) and (min) refer to the *maximum and minimum* values expected. Symbols otherwise refer to Figure 15-7(a).

Section 15.5 / DC REFERENCE SUPPLIES

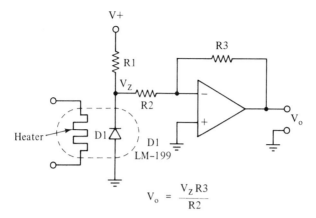

FIGURE 15-2. Stabilized voltage reference supply.

thermal environment. In the past this meant a component oven, but now "four-terminal" zeners are available that use a heating element and zener diode thermally coupled together on the same IC substrate.

The manufacturer of the LM-199 devices uses a zener diode nested among the transistors of a class-A amplifier, which produces constant heat dissipation under zero-signal conditions.

The circuit in Figure 15-2 uses a temperature-controlled zener diode to set the potential at one input of an operational amplifier. You should be able to understand the operation of this circuit once you complete studying Chapter 12.

The voltage at the output terminal of the circuit in Figure 15-2 is given by the normal transfer function of the inverting follower operational amplifier:

$$V_o = -V_z \times \frac{R3}{R2} \qquad (15\text{-}1)$$

The zener voltage and the desired output voltage are usually known, so we will need to rearrange Equation (15-1) to find either the ratio $R3/R2$ or, if one or the other resistor is selected for convenience sake, the value of the remaining resistor. Also, note that this circuit produces a negative output voltage, so it can be used only with data converters that either need or allow a negative reference source (e.g., DAC-08 or MC1408-8 devices).

The circuit for a positive output operational amplifier reference source is shown in Figure 15-3. In this case, the operational amplifier is connected in the noninverting follower configuration. The zener reference

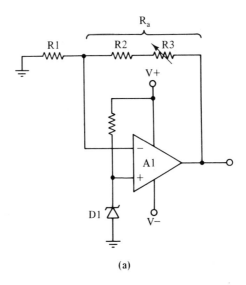

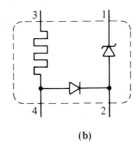

FIGURE 15-3. (a) Precision reference supply. (b) Temperature stabilized Zener LM199.

diode (which could also be an LM-199 or similar device) is connected with its series current-limiting resistor to the noninverting input of the amplifier. This connection makes the input voltage of the amplifier equal to the zener potential. The transfer equation for this circuit is

$$V_o = V_z \times \left[\frac{R_a}{R1} + 1\right] \quad (15\text{-}2)$$

where R_a is the series combination $R2 + R3$, and the other factors bear their normal meanings.

The potentiometer in the circuit of Figure 15-3 is used to trim the output voltage to a precise value. For this reason, it should have a max-

Section 15.5 / DC REFERENCE SUPPLIES

imum value of approximately 10 percent of R_a. Further, the potentiometer should be a 10 to 20 turn "trimpot" type in order to increase the resolution per turn. All of the resistors in this circuit should be precision types of one percent or better tolerance. The purpose in selecting precision resistors is not as much for the precise setting of the amplifier gain (the potentiometer does that job), but to acquire low temperature coefficient resistors. It does little good to specify a 20-turn potentiometer for $R3$, only to use carbon composition resistors for $R1$ and $R2$, thereby introducing more drift than the potentiometer could compensate. The low temperature coefficient resistors will keep the gain of the amplifier reasonably stable even after warm-up of the circuit.

In some applications, the circuits of Figures 15-2 and 15-3 can use an ordinary zener diode, especially if the ambient temperature remains relatively constant. But for most instrumentation and control applications, a premium reference diode must be used in place of the zener. Recommended are the LM-199, which keeps the buried zener at a constant temperature or one of the band-gap zener reference diodes. The LM-199 (and the lower grade LM-299 and LM-399 devices in the same series) offer good temperature-tracking specifications because the zener element is buried on the same semiconductor die with a class-A amplifier that has its input terminals shorted together. The effect of this configuration is that the class-A amplifier transistors dissipate a constant amount of power, thus producing a constant temperature environment on the IC die. Burying the zener also produces lower noise operation, and provides a 20 ppm long term stability, and short term stability of 1 ppm. The rated terminal voltage of the LM-199 is ±2 percent, but this is corrected by the operation of potentiometer $R3$ in Figure 15-3.

The operation of the LM-199 (and others of the series) is such that the diode-side connections are identical to zener diode connections. The heater terminals, however, must be connected to a DC source of 9 to 40 volts. A diagram of the LM-199 circuit is shown in Figure 15-3(b).

An example of a band-gap zener diode is shown in Figure 15-4. This device is the *Ferranti Semiconductor, Inc.* ZN-458A/B device, but it is representative of the products of several manufacturers. The band-gap zener diode is on the same substrate with an amplifier and output transistor that will allow the device to pass up to 120 milliamperes of current. Note, however, that best performance occurs when the diode is operated at a lower current, in the 2 to 15 mA range. The Ferranti devices are available in several different voltages and with different stability ratings. Most of the devices in this line offer either 1.26 volts or 2.45 volts, with stability ratings to 10 ppm, at a temperature coefficient of 0.003 percent per degree centigrade. The series current-limiting resistor is selected according to the same rules as those for ordinary zener diode regulators.

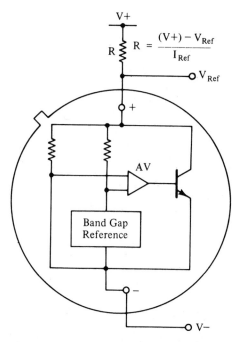

FIGURE 15-4. Ferranti band-gap voltage reference IC.

A final example of the band-gap reference diode is shown in Figure 15-5, the Intersil ICL8069 device. The specifications for various grades of this diode are shown in Figure 15-5(b), while the case is shown in Figure 15-5(c). The device is available in both commercial (0°C to +70°C) and military (−55°C to +125°C) temperature ranges, with temperature coefficients from 0.001 %/°C to 0.01 %/°C.

Circuits using the ICL8069 device are shown in Figure 15-6. The circuit in Figure 15-6(a) shows the use of an operational amplifier with the ICL8069 to produce a 10.000 volt reference source. In Figure 15-6(a), the reference diode is connected in the negative feedback path of the operational amplifier, while a resistor feedback network is used to the noninverting input. The potentiometer in this circuit is selected to have a value of approximately 10 percent of the total, and should be a 10 to 20 turn type. All resistors in this circuit should be precision types, again to take advantage of the temperature coefficient.

The operational amplifier selected for a reference source should be able to produce low drift characteristics. In general, the lowest cost operational amplifiers, such as the 741 device, are not suitable. A premium de-

Section 15.5 / DC REFERENCE SUPPLIES

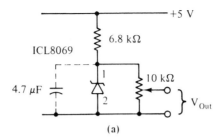

Max Temperature Coefficient of V_{Ref}	Temp Range	Order Part #
0.001%/°C	0°C to +70°C	ICL8069ACQ
0.0025%/°C	0°C to +70°C	ICL8069BCQ
0.005%/°C	−55°C to +125°C	ICL8069CMQ
0.005%/°C	0°C to +70°C	ICL8069CCQ
0.01%/°C	−55°C to +125°C	ICL8069DMQ
0.01%/°C	0°C to +70°C	ICL8069DCQ

(b)

(c)

FIGURE 15-5. Intersil ICL8069 voltage reference.

vice such as the LM108, LM156, or CA3140 are required. It makes little sense to specify a premium reference diode, and then connect it into a circuit in which the drift factor and offset potentials/currents negate the advantages of the reference diode.

In Figure 15-6(b) we see the use of the ICL8069 device in a data converter circuit without the benefit of the operational amplifier. In this case, a series combination of a fixed resistor and a potentiometer is used to trim

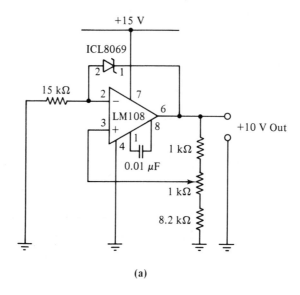

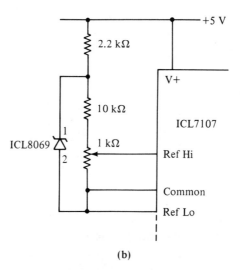

FIGURE 15-6. Using the ICL8069 (a) precision reference source. (b) with ICL7107 ADC.

the fixed output of the ICL8069. Note that the drift/stability factors of these diodes are quite good, but the actual terminal voltage is nominal, so it may be as much as two percent in error. The potentiometer allows us to cancel some of that error.

15.6 INTEGRATED CIRCUIT REFERENCE SOURCES

Integrated circuit technology has allowed the development of a large number of special function devices. One device that takes good advantage of the technology is the voltage reference. Several companies now offer voltage reference devices that are suitable for use in data converter applications. Although there are several different devices on the market, we will cover only one of the more popular: the REF-01/02 by *Precision Monolithics, Inc.*, shown in Figure 15-7. The REF-01 is a 10.000 volt device (Figure 15-7(b)), while the REF-02 is a 5.000 volt device (Figure 15-7(b)). The REF-02 also offers a temperature output (pin no. 3 in Figure

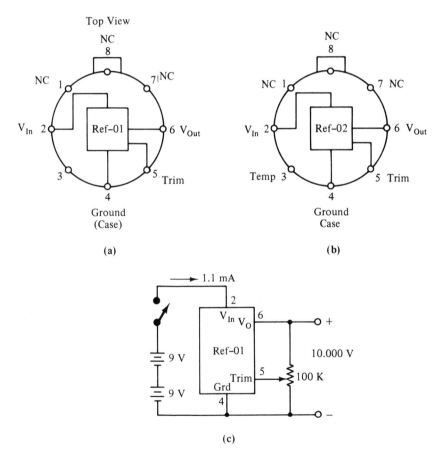

FIGURE 15-7. (a) REF-01 voltage reference source. (b) REF-02 voltage reference source. (c) Connections for the REF-01 and -02.

15-7(b)) voltage that produces a slope of 2.1 mV/°C. This feature makes the REF-02 useful in temperature measurement applications and also provides a signal that can be used to trim circuits according to temperature changes.

The basic circuit of the REF-01 and REF-02 is shown in Figure 15-7(c). A sample of the output voltage is fed back to the device through a potentiometer that can be used to trim the output voltage to within a few millivolts of the rated output voltage. These devices will sink up to 20 mA of current, but one is advised to use as little current as possible in reference sources.

15.7 CURRENT REFERENCE SOURCES

Some data converter circuits require a current reference instead of a voltage reference. In most of those cases, we can create a current reference by using one of the available reference voltage sources and a resistor; Ohm's law has not yet been repealed by Congress! The simple tactic is to apply a reference voltage to one end of a precision resistor connected to the reference current input of the DAC. The reference current I_{ref} is then simply the quotient of the reference voltage over the resistance. But sometimes it is also desirable to use a separate current reference source, and we have several methods for accomplishing this job.

One way to provide a current reference is to purchase a reference current "diode." These devices will be rated at some nominal current, and produce relatively consistent results. But the device is actually little more than a junction field effect transistor (JFET) that is connected with the gate and source terminals shorted together. Although the connection is made internally, making the device look like a two-element "diode," it is actually similar to Figure 15-8(a). All JFETs have a certain knee current at which the device becomes saturated. Applied voltages higher than that required to saturate the device will not produce a higher source-drain current. We can, therefore, use the JFET in that region to produce a constant reference current. A method for making the current variable over the range 6 μA to 2 mA is shown in Figure 15-8(b). Here we have a potentiometer connected as a rheostat in series with the source of the JFET, while the gate continues to be connected to the source.

The use of a pair of bipolar NPN transistors as a constant current source is shown in Figure 15-8(c). This circuit uses slight regeneration to servo out changes on the load resistance, hence keeping the load current $I1$ constant. The derivation of the equation for $I1$ is left as an exercise for the reader.

Section 15.7 / CURRENT REFERENCE SOURCES

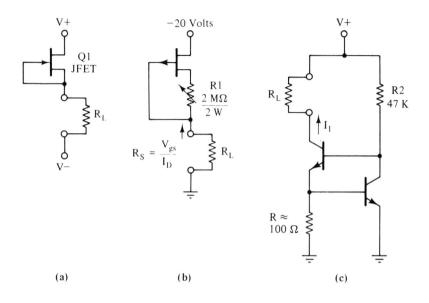

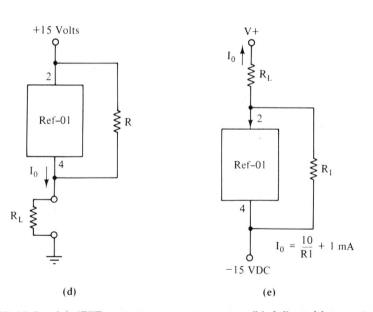

FIGURE 15-8. (a) JFET constant current source. (b) Adjustable constant current source. (c) Bipolar constant current source. (d) REF-01 constant current source. (e) REF-01 constant current sink.

The REF-01 and REF-02 devices can be used as a constant current source by using the circuits of Figures 15-8(d) and 15-8(e). The circuit in Figure 15-8(d) is a current source, while that of Figure 15-8(e). is a current sink. In both cases, the output current is given by

$$I_{out} = \frac{10.0\ V}{R} + 1\ mA \qquad (15\text{-}3)$$

where:

I_{out} is the current in milliamperes

R is the resistance in kiloohms

V is the potential 10.000 volts

16
Interfacing Keyboards, Switches, and Displays

16.1 OBJECTIVES

1. To learn to interface ordinary ASCII keyboards to the microcomputer.
2. To learn to interface pushbutton switches to the microcomputer, thereby allowing special functions and custom keyboards.
3. To adapt keyboards to specific microcomputers.
4. To learn to interface seven-segment LED numeric displays to the microcomputer.

16.2 SELF-EVALUATION QUESTIONS

Before studying the material in this chapter, try answering the questions given below. These questions test your knowledge of the subject matter. If you cannot answer a particular question, then place a check mark beside it and look for the answer as you read the text.

1. How may a too-short strobe pulse be lengthened to accommodate the cycle-time of a microcomputer?
2. Which signals from a Z80 microprocessor are needed to generate (a) an OUT signal, and (b) an IN signal.
3. How may a transient strobe signal be generated from a keyboard that produces a constant level on the strobe bit as long as the key is depressed by the operator?
4. Draw a circuit for interfacing up to eight individual pushbuttons to a microcomputer data bus or input port.

223

16.3 KEYBOARDS

The ASCII keyboard is probably the most common method for talking to a microcomputer. On simpler computers we might find a calculatorlike keyboard that will allow us to input hexadecimal numbers for direct machine language programming or data entry. In both cases we will need to know the methods for interfacing the keyboard, and the programming needed to accommodate the keyboard (the computer does not "naturally" understand the keyboard).

Figure 16-1 shows the basic circuit for a typical keyboard. This particular type is for the ASCII code, but the circuit is similar for hexadecimal keyboards (which would use a smaller keyboard X-Y matrix and probably a different IC).

Device IC1 is a keyboard integrated circuit. It is basically a *read only memory* (ROM) in which the binary codes for the various ASCII codes are stored in address locations that are uniquely accessed by single contact closures on the X-Y matrix.

The actual keyboard is basically a crosspoint switch in which depressing a given key will short together a unique combination of X and Y terminals on IC1. When a key is depressed, then, the binary code (representing an ASCII character) is output to lines B1 through B7.

The keyboard contains two additional switches, S1 and S2. These switches are for the control (CNTL) and shift (SHFT) functions, respectively.

A strobe pulse or signal is used to let the outside world know when the data is new and valid. In some designs, there will be a trashy signal on the output lines except when a pushbutton is pressed. The computer must know when this data is a valid ASCII character, and when it represents nothing more than random noise. In still other cases, the output will be latched, so the computer needs to know when the data is truly new and when it is the old data that has already been input. Two types of strobe signal are used, *transient* and *level*. The transient form of strobe pulse (Figure 16-2(b)) is a short-duration pulse that is issued once per keystroke. When the data on the keyboard output lines becomes stable and valid, then a brief pulse is generated. The time between the closure of the switch and the production of the output pulse may be only nanoseconds. The level form of strobe signal is shown in Figure 16-2(a). This signal is a constant voltage level that remains as long as the keyboard operator is depressing a key. The level type of strobe signal is seen in low cost keyboards and in those for certain special purpose applications. In Figure 16-1, the level type of strobe signal is available at the output of the NAND gate (IC3), while the transient form is available at the output of IC2 (a

Section 16.3 / KEYBOARDS

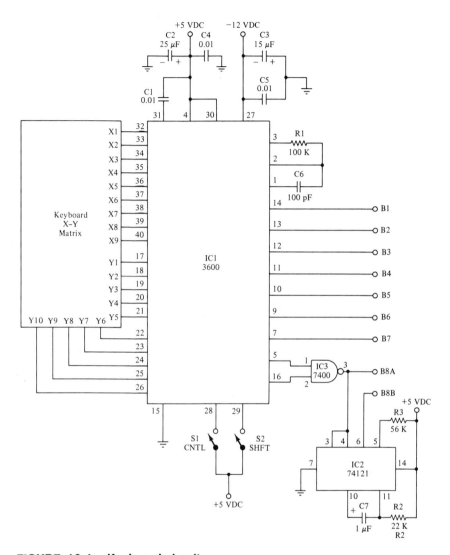

FIGURE 16-1. Keyboard circuit.

one-shot). In an eight-bit microcomputer, we can use the lower order seven bits of the data bus or input port for the ASCII data, and the highest order bit to carry the strobe signal. In Figure 16-1, the level strobe signal is available as bit B8A, while the transient form is at bit B7B. Of course, in any given application, only one of these will be used.

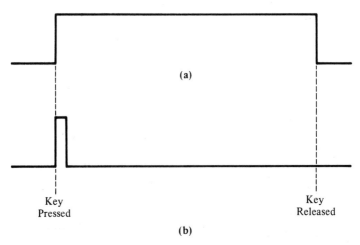

FIGURE 16-2. Two types of strobe signal. (a) Constant level. (b) Pulse.

16.4 INTERFACING KEYBOARDS TO I/O PORTS

Most currently available microcomputer keyboards have TTL-compatible output lines. Similarly, most modern microcomputers have TTL-compatible input ports. We can, therefore, directly connect the output lines of the keyboard to the input lines of the computer. Furthermore, the seven data bits plus strobe bits are easily compatible with the eight-bit format of the typical microcomputer input port. It is merely necessary to connect the lines of the outputs to the computer inputs.

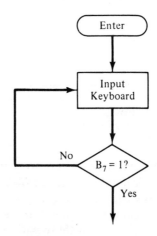

FIGURE 16-3. Keyboard input routine.

We must, however, write a simple program to service the keyboard. It is the usual practice of designers to have the microcomputer loop endlessly while looking for a valid strobe bit. An example is shown in Figure 16-3. Here we are inputting the binary word from the input port (i.e., the keyboard data). Bit B7 of the input port (B8 of the keyboard—they are often numbered slightly differently) will be logical-1 (HIGH) when the data is valid. We will, therefore, want to mask all bits but B7, and then test for zero or nonzero. For example, we can AND the input word with 10000000, which will set all bits to zero except B7 when B7 of the input signal is also 1. In a typical program, the computer will branch back to the input instruction on result = 0, and to the program for storing the input data on result = 1.

16.5 INTERFACING KEYBOARDS TO DATA BUS

The data bus in a microcomputer is used for many different purposes. The use of the bus changes virtually every time a new instruction is either fetched or executed—in other words once every cycle. We must not, therefore, allow any one device to command the data bus all of the time. In the case of a keyboard or similar peripheral we can use a circuit similar to Figure 16-4 to interface to the data bus.

The device used to connect the keyboard to the data bus is a tri-state driver. This IC has a third possible output state that exists when pin no. 1 is HIGH: high impedance to both +5 volts and ground. In this third state the internal circuitry is effectively disconnected from the output pins. Pin no. 1 controls the operation of the IC. When pin no. 1 is HIGH, then the device floats across the data bus at high impedance, so it cannot affect the bus. If, on the other hand, the level applied to pin no. 1 is LOW, then the outputs of the internal buffer amplifiers are connected to the data bus. If the $\overline{\text{IN}}$ signal drops LOW when the keyboard data is valid, then data will be input to the computer.

Notice that we show two alternate methods for connecting the strobe signal to the computer. In one case, we would apply the strobe to the highest order bit of the 8212 input port device. In that case, we would write a loop program (as described earlier), and hope that it works properly. But there is also another method that is not so wasteful of CPU time. We would allow the microcomputer to perform other chores unless the strobe signal causes an interrupt. This is accomplished by connecting the strobe output of the keyboard to the interrupt line of the computer (inverting polarity is sometimes needed). When the computer senses that the interrupt line is LOW, it will stop executing the program that is in

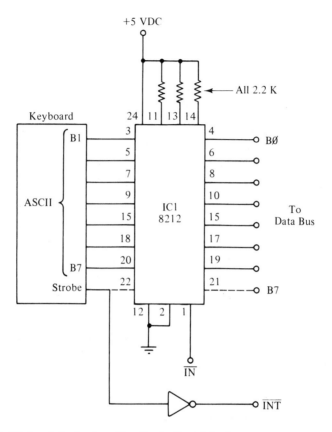

FIGURE 16-4. Interfacing directly to the data bus.

progress and jump to a subroutine specified by the interrupt. In this case, the subroutine would have to be a keyboard input program.

In an earlier chapter we discussed the generation of device select pulses and decoder circuits. Let's review how the IN and OUT signals are generated. Interfacing the keyboard to the data bus required an $\overline{\text{IN}}$ signal, i.e., a signal that will drop LOW when the computer wants to input keyboard data. This signal is shown for the Z80 based machines in Figure 16-5. For the Z80 there are three conditions that must be met before an input operation takes place: (1) I/O request line ($\overline{\text{IORQ}}$) goes LOW, (2) read line ($\overline{\text{RD}}$) goes LOW, and (3) the correct input port address is present on the lower eight bits of the address bus. In the circuit of Figure 16-5 the 7430 device is used as the address decoder; its output will drop LOW only when the correct address (in this example 11010011) is present on the address bus. A 7442 BCD-to-1-of-10 decoder is used to detect when all

Section 16.5 / INTERFACING KEYBOARDS TO DATA BUS

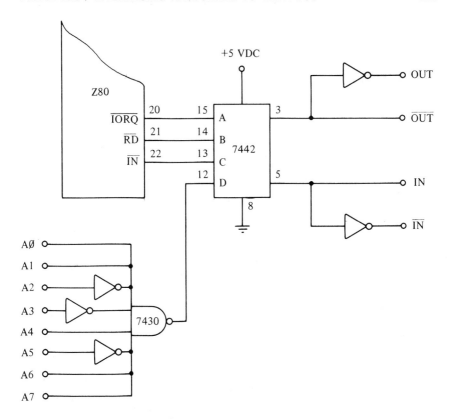

FIGURE 16-5. Generating select pulses.

three conditions are simultaneously present. This device will examine a BCD word applied to its inputs and then cause the correct decimal output to drop LOW. The way we have the 7442 connected requires that the binary word 0100 be applied to the BCD inputs before pin no. 5 (i.e., the "4" output) goes LOW. When that occurs we have our $\overline{\text{IN}}$ signal.

Solving Some Keyboard Interface Problems

It is often necessary to interface a keyboard to a computer that is not totally compatible with the computer. For this type of operation we will usually need to solve some practical problems. In an earlier chapter we discussed the problem of interfacing different logic families; that information will not be repeated here. We will, however, address some other common problems.

A keyboard input subroutine requires a certain amount of time to execute. In some cases, for example, we might find that the program requires as many as 15 machine cycles (at 1 microsecond each), requiring a total of 15 μS for execution. If the keyboard strobe signal does not stay active for at least that length of time, then it is likely to be missed as the program loops through its cycle. Several different alternatives are open to us, two of which are shown in Figures 16-6 and 16-7.

In Figure 16-6 we see a method for stretching the strobe pulse by using a 74121 monostable multivibrator. Such a circuit will produce one output pulse of constant duration for each input trigger pulse received. We can set the duration of the output pulse by adjusting the values of R1 and C1 according to the equation

$$T \approx 0.69 \, R_1 \, C_1$$

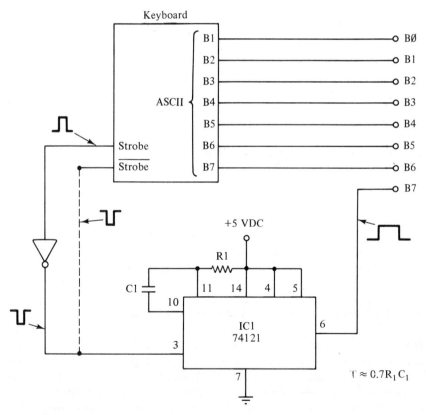

FIGURE 16-6. Stretching the too-short pulse.

Section 16.5 / INTERFACING KEYBOARDS TO DATA BUS

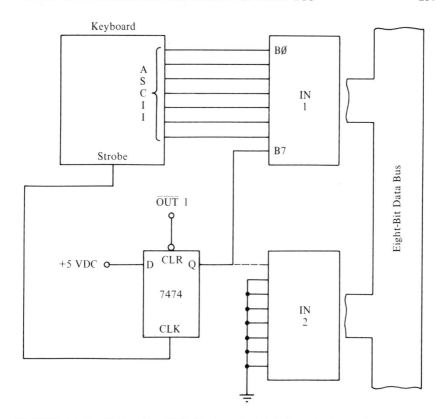

FIGURE 16-7. Using the 7474 flip-flop to latch the strobe pulse.

In this case we are not really "stretching" the pulse, but rather we are substituting a longer pulse for the short strobe pulse. This circuit, or its alternative, is called for any time the strobe pulse is too short for the computer's response time.

The second alternative, shown in Figure 16-7, uses a Type-D flip-flop to hold the strobe pulse after it is generated. Notice that the *strobe* output of the keyboard is applied to the *clock* terminal of the flip-flop. When the strobe pulse arrives, a HIGH is transferred to the Q output of the FF because the D-input is connected permanently HIGH. The data outputs of the keyboard are connected to the computer data bus through a standard input port (IN1). We again have an alternative; we could either connect the Q output of the FF to bit B7 (the MSB) of IN1, or we could connect it to one bit of a second input port (IN2, in this case). The first choice is the least wasteful of resources. Normally, when the data is not valid, the MSB of IN1 will be LOW. The computer will execute a loop that seeks the condition B7 = HIGH before it executes the input and store program. Following storage of the ASCII keyboard data somewhere in memory, the computer will execute a dummy output instruction to port no. 1 (i.e., generate an $\overline{OUT1}$ device select pulse) that is used to clear the FF, making Q = LOW again until a new strobe pulse indicates that the operator has hit another key.

Some keyboards, especially those found as industrial surplus or among the very cheapest hobbyist grade models, will not have latched outputs. The data on the outputs of these models will disappear as soon as the operator lifts a finger off the key. In some cases, it is preferred that the data remain valid until the next valid data is sent to the computer. In order to add a latch to a keyboard we need an IC such as the 74100 shown in Figure 16-8. This IC device is a dual four-bit data latch. By connecting the two strobe inputs together (i.e., short S1 and S2 together), we can make an eight-bit data latch. We will ground one input of the 74100 in order to keep its output continually LOW. The ASCII data lines from the keyboard are connected to the remaining seven input lines. The output lines are connected to the input port of a microcomputer. The strobe line from the keyboard is connected to the strobe inputs of the 74100. When the strobe pulse comes along, indicating that the ASCII data applied to the 74100 inputs is now valid, the 74100 will transfer the input data to the output lines B0-B7. After the strobe pulse has vanished, the ASCII data will remain on the 74100 outputs.

Another problem is the nature of the strobe signal itself. Recall that the strobe signal might be either a level or a pulse. In some cases, it is better to use a pulse-type strobe signal than a level. This becomes espe-

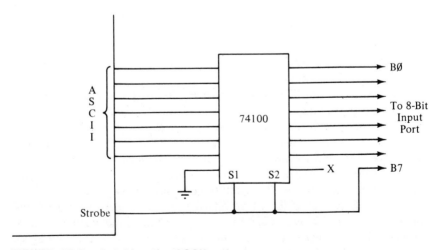

FIGURE 16-8. Latching the ASCII output.

Section 16.6 / INTERFACING PUSHBUTTONS TO THE MICROPROCESSOR

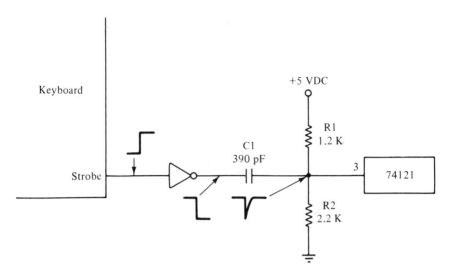

FIGURE 16-9. Generating a pulse strobe from a constant level strobe.

cially necessary when the software being executed by the computer looks for the disappearance of the strobe signal to do some chore (it happens!). The circuit in Figure 16-9 can help this problem. If the strobe signal is a positive level, then the output of the inverter will be a negative-going positive signal. This signal is differentiated in the R1/R2/C1 network, and the resultant negative-going spike is used to trigger a 74121 monostable multivibrator. If the strobe signal from the keyboard is already negative-going, then delete the inverter. In some cases, it might be desirable also to include a diode connected in the proper direction across one of the resistors (usually R1) so that the positive spike from the differentiator cannot work any mischief in the circuit.

16.6 INTERFACING PUSHBUTTONS TO THE MICROPROCESSOR

Pushbuttons might be interfaced to a microcomputer/microprocessor for any of several different reasons. Among them might be a special purpose of limited use keyboard. We find several different kinds of electrical switch in electronic instruments. We know the different forms (for example, SPST, SPDT, DPST, DPDT, and so forth). We also find alternate action switches. These will perform one operation on the first press and then the opposite operation on the second press. For example, we might find an SPST "AA" switch that will close on the first press and then open on the second press. All of these switches can be simulated using simple switches and some software in a microcomputer. For example, consider a simple situation such as that shown in Figure 16-10. The actual switch is a simple SPST type and may be either a toggle switch or a pushbutton (it

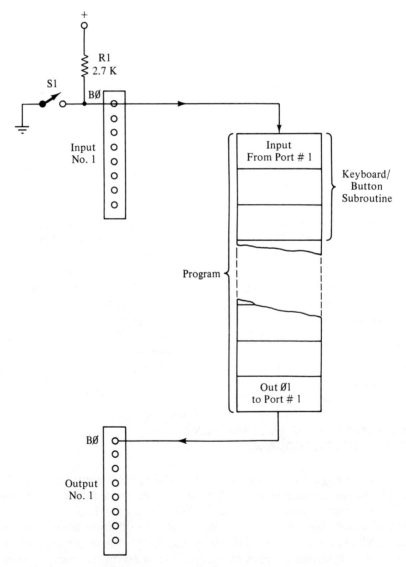

FIGURE 16-10. Using a direct-interface switch.

Section 16.6 / INTERFACING PUSHBUTTONS TO THE MICROPROCESSOR

can also be an electronic switch). There is a pull-up resistor between the junction of the switch and port bit B0 and the +5 volt line. This means that B0 will be HIGH when S1 is open, and LOW when S1 is closed. Let's say that the desired action is to have the computer write a HIGH to bit B0 of output port no. 1 whenever the switch is closed (i.e., B0 of input port no. 1 is LOW). We would write a short program that would continuously monitor port no. 1 and determine when B0 is LOW. It would then write a data word with B0 HIGH (i.e., 01H, 07H, FFH, etc.) to output port no. 1. Of course, it is not necessary that the program perform this specific operation. Just as often, the computer would use the data provided for some other purpose. It is common on microprocessor-controlled instruments to use one input port to inform the computer of certain desired conditions. In that case, all eight bits of the selected input port might have a switch connected. The operator could then program the computer for whatever conditions are existing or, possibly, desired.

Figure 16-11 shows two methods for interfacing switches to the data bus of the computer without the need for existing input ports. Of course, this operation requires us to *make* an impromptu input port for the switch involved.

The circuit in Figure 16-11(a) shows how to interface with a single switch. We require an input buffer (either inverting or noninverting will suffice, but changes will be required in the software) that has a tri-state output terminal; the 74125 and 74126 are examples. In addition, the *chip enable* terminals must be independent of other sections of the chip. When the $\overline{CE}$ terminal is brought LOW by the generation of an $\overline{IN1}$ device select pulse, the data at the input of the tri-state buffer is transferred to the output, hence to the data bus of the microcomputer. We could use either of two approaches shown for connecting a switch to the input of the buffer. In the case of S1, we use the same sort of circuit as in Figure 16-10; the pull-up resistor is needed for S1 but not if the alternate method (using the FF) is used. Alternatively, we could use a type-D flip-flop to hold the data after the operator releases the button. We would want to use this method when (a) the computer does not cycle back to the input port containing the switch fast enough to not miss an operation, or (b) when the computer will periodically interrogate the switch to find if some preset condition still exists. In some cases, for example, the intent will be to have the computer continuously execute the program until someone resets the output of the FF. In the case shown, we can reset the FF either by a software command (i.e., execute a dummy output to generate an $\overline{OUT1}$ signal) or by a hardware pushbutton (S3 in Figure 16-11(a)).

The other circuit is shown in Figure 16-11(b). In this case, we are doing essentially the same thing as in Figure 16-10, except that we are required to create our own input port. The switches are connected to the

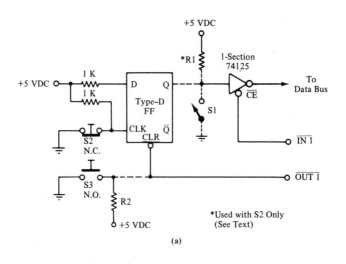

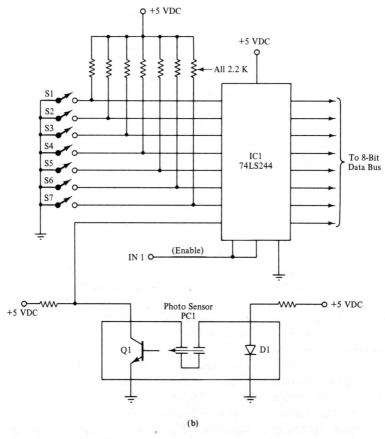

FIGURE 16-11(a). Switch. **(b).** Interfacing switches to data bus.

Section 16.6 / INTERFACING PUSHBUTTONS TO THE MICROPROCESSOR   237

input of an eight-bit noninverting buffer that has tri-state outputs. When the two strobe lines are brought HIGH by generation of the IN1 signal, the data on the respective input lines are transferred to the output lines, hence to the eight-bit data bus of the microcomputer. The software of the computer will then sort out the respective meanings of the various open and closed conditions for the switches.

Note that this circuit also includes an electronic photo switch. These devices are often used in mechanical instruments to indicate some external condition. The photo switch consists of a light emitting diode (LED) and a phototransistor. These components are mounted inside of a light-tight housing that has an optical path that turns on the transistor when the LED is illuminated. In some, the entire photo switch is enclosed, and these are called optoisolators. In still other cases, there will be a slot between the transistor and the LED to all or some external device to blind the transistor. We see this type of circuit in instruments such as printers. There will be a metal or plastic tab on the print-head carriage that will be inserted into the slot when the head carriage reaches the end of its travel. This arrangement allows us to generate a HIGH when the head carriage is at the end of travel, and a LOW all other times.

An Alternate-Action Circuit

Microprocessor-based instruments sometimes require an alternate-action (AA) switch. Sometimes this is a matter of necessity; sometimes it is a matter of convenience or front panel design. An alternate action switch is one that performs opposite actions on successive closures of the contacts. Mechanical AA switches can be quite complicated when compared with a simple SPST switch. In this section we will discuss a method for making an SPDT AA switch using a simple, normally closed (NC) SPST pushbutton switch and a dual Type-D flip-flop. The typical TTL Type-D FF requires a LOW on the CLR line for the FF to be reset (i.e., made Q = LOW, and NOT-Q = HIGH). When the power is first applied to the instrument, a *power-on reset* pulse is generated; this pulse is coupled through NOR gate G1 to the CLR lines of the flip-flops. This action will initialize the FFs to the condition that makes output A = LOW and output $\overline{A}$ = HIGH.

Recall the action of the Type-D flip-flop. When the clock input is made HIGH, the data on the D input will be transferred to the Q output. In the circuit of Figure 16-12, the D input of FF1 is connected permanently HIGH. The normally closed pushbutton is connected to the CLK (clock) input, along with a pull-up resistor to +5 volts DC. When the button is pressed, the switch contacts open, and this brings the CLK input

238 INTERFACING KEYBOARDS, SWITCHES, AND DISPLAYS / Chapter 16

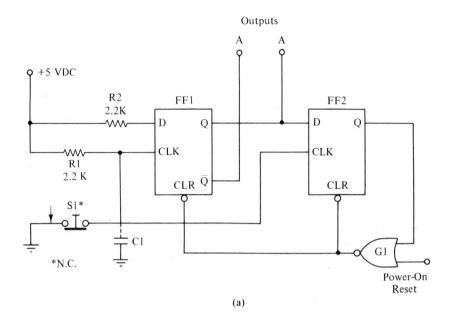

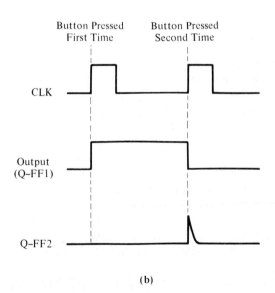

FIGURE 16-12(a). Alternate action switch. **(b).** Timing diagram.

HIGH, thereby transferring the HIGH on the D input to the $\overline{Q}$ output. At this time, the A output of the circuit is made HIGH and the $\overline{A}$ output of the circuit is LOW. When the button is pressed a second time, we find that the D input of FF2 is now HIGH (it had been LOW before), so the Q output of FF2 goes HIGH. This level is coupled back to the CLR line of FF1 and FF2 through gate G1 (the same as for the power-on reset pulse). This has the effect of resetting the flip-flops, making both Q output LOW. At this time, the A output of the switch goes LOW and the $\overline{A}$ goes HIGH; alternate action! The timing diagram for this circuit is shown in Figure 16-12(b).

Notice that an optional capacitor is used in this circuit. Its purpose is to suppress pushbutton contact bounce. Capacitor C1 must have a value sufficient to suppress the contact bounce pulses, but not so high as to interfere with the operation of the circuit; less than 500 pF should be sufficient in most cases.

16.7 SUMMARY

1. In some cases, ASCII keyboards can be interfaced directly to a microcomputer input port; in other cases a special input port must be created.
2. Problems such as the strobe pulse being too short or inappropriate must sometimes be solved for successful interfacing.
3. Specialized keyboards or individual switches can be interfaced either directly to the microcomputer via an existing I/O port or a special port must be created.

16.8 QUESTIONS

16-1. Both ASCII and Hexadecimal keyboards may use an _____ matrix in order to reduce the number of connections needed between the pushbuttons and the encoding circuitry.

16-2. A _____ pulse indicates to the outside world when the data on the keyboard output is valid.

16-3. Two types of strobe signal are _____ and _____.

16-4. Most currently available keyboards use _____ levels for the data signals.

16-5. ASCII encoding requires _____ data bits and one _____ bit.

16-6. Describe how an ASCII or Hex keyboard may be connected to a microprocessor data bus.

17
Basics of Data Conversion—D/A

17.1 OBJECTIVES

1. To list the two major types of Digital-to-Analog Converter (DAC).
2. To describe the operation of the principal types of DAC.
3. To list principal applications of DACs.
4. To calculate DAC output voltage in practical circuits.
5. To calculate DAC resolution in practical circuits.

17.2 SELF-EVALUATION QUESTIONS

Before studying the material in this chapter, try answering the questions given below. These questions test your knowledge of the subject. If you cannot answer any particular question, then place a check mark beside it, and then look for the answer as you read the text.

1. Most binary DACs use either the _____ _____ or _____ resistance ladders.
2. What is a multiplying DAC?
3. How many different *states* can be represented by an eight-bit DAC?
4. What is the greatest decimal number that can be represented in straight binary by an eight-bit DAC?
5. Each step in a 10-bit DAC with a full output potential of 10.00 volts represents an output voltage change of _____ mV.

17.3 WHAT ARE DATA CONVERTERS?

Analog circuits and digital instruments occupy mutually exclusive realms. In the analog world, a signal may vary between upper and low limits, and

Section 17.4 / BINARY RESISTANCE LADDER CIRCUITS

may assume any value within the range. Analog signals are continuous between limits. But the signals in digital circuits may assume only one of two discrete voltage levels, i.e., one each for the two binary digits, 0 and 1. A *data converter* is a circuit or device that examines signals from one of these realms and then converts it to a proportional signal from the other.

A *digital-to-analog converter* (DAC), for example, converts a digital (i.e., binary) "word" consisting of a certain number of *bits* into a *voltage* or *current* that represents the binary number value of the digital word. An eight-bit DAC, for example, may produce an output signal of 0 volts when the binary word applied to its digital inputs is 00000000_2, and (say), 2.56 volts when the digital inputs see a word of 11111111_2. For binary words applied to the inputs, then, a porportional output voltage is created.

In the case of an *analog-to-digital converter* (ADC, or A/D), an analog voltage or current produces a proportional binary word output. If an 8-bit ADC has a 0- to 2.56-volt input signal range, then 0 volt input could produce an output word of 00000000_2, while $+2.56$-volt level seen at the input would produce an output word of 11111111_2.

Data converters are used primarily to interface transducers (most of which produce analog output signals) to digital instruments or computer inputs, and to interface digital instrument outputs to analog-world devices such as meter movements, chart recorders, motors, etc.

17.4 BINARY RESISTANCE LADDER CIRCUITS

Figure 17-1 shows a *binary weighted resistance ladder* and operational amplifer used as a binary DAC. The operation of this circuit can be deduced from operational amplifier theory given in Chapter 12.

The resistors in the ladder are said to be binary weighted, because their values are related to each other by powers of two. If the lowest value resistor is given the value R, then the next in the sequence will have a value $2R$, followed by $4R$, $8R$, $16R$, all the way up to the nth resistor (last one in the chain), which has a value of $(2^{(n-1)})R$.

The switches ($B1$ through B_n) represent the input bits of the digital word. Although shown here as mechanical switches, they would be transistor switches in actual practice. The switches are used to connect the input resistors either to ground, or voltage source E, to represent binary states 0 and 1, respectively. Switches $B1$ through B_n create currents $I1$ through I_n, respectively, when they are set to the "1" position.

We know from Ohm's law that each current $I1$ through I_n is equal to the quotient of E and the value of the associated resistor, i.e.:

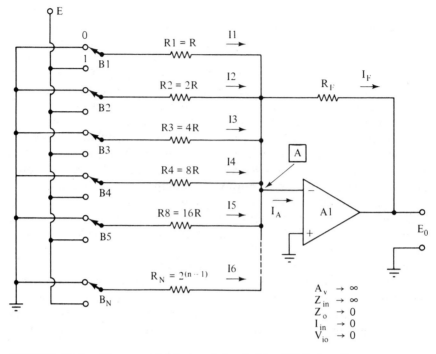

FIGURE 17-1. Binary weighted resistor ladder DAC circuit.

$$I1 = E/R1 = E/R$$
$$I2 = E/R2 = E/2R$$
$$I3 = E/R3 = E/4R$$

$$\cdot \qquad \cdot \qquad \cdot$$
$$\cdot \qquad \cdot \qquad \cdot$$
$$\cdot \qquad \cdot \qquad \cdot$$

$$I_n = E/R_n = E/(2^{(n-1)})R$$

The total current into the junction (point A in Figure 17-1) is expressed by the summation of currents $I1$ through I_n:

$$I_A = \sum_{i=1}^{n} \frac{a_i E}{2^{(i-1)} R} \qquad (17\text{-}1)$$

where I_A = the current into the junction (point A) in amperes (A)
E = the reference potential in volts (V)

Section 17.4 / BINARY RESISTANCE LADDER CIRCUITS

R = the resistance of $R1$ in ohms (Ω)

a_i = either "1" or "0," depending upon whether the input bit is "1" or "0"

n = the number of bits, i.e., the number of switches

From operational amplifer theory we know that

$$I_A = -I_f \tag{17-2}$$

and
$$E_0 = I_f R_f \tag{17-3}$$

So, by substituting Equation (17-2) into Equation (17-3), we obtain

$$E_0 = -I_A R_f \tag{17-4}$$

and substituting Equation (17-1) into Equation (17-4) yields

$$E_0 = -R_f \sum_{i=1}^{n} \frac{a_i E}{2^{(i-1)} R} \tag{17-5}$$

Since E and R are constants, we usually write Equation (17-5) in the form

$$E_0 = \frac{-ER_f}{R} \sum \frac{a_i}{2^{(i-1)} R} \tag{17-6}$$

Example 17-1

A four-bit (i.e., $n = 4$) DAC using a binary weighted resistor ladder has a reference source of 10 volts dc, and $R_f = R$. Find the output voltage E_0 for the input word 1011_2. (*Hint:* for input 1011, $a_1 = 1$, $a_2 = 0$, $a_3 = 1$, and $a_4 = 1$.)

Solution

$$E_0 = \frac{-ER_f}{R} \sum \frac{a_i}{2^{(i-1)} R} \tag{17-6}$$

$$E_0 = \left[\frac{(-10 \text{ V})(R)}{(R)}\right] \left[\frac{1}{2^{(1-1)}} + \frac{0}{2^{(2-1)}} + \frac{1}{2^{(3-1)}} + \frac{1}{2^{(4-1)}}\right]$$

$$E_0 = (-10 \text{ V}) \left[\frac{1}{2^0} + \frac{1}{2^2} + \frac{1}{2^3}\right]$$

$$E_0 = (-10 \text{ V}) \left[\frac{1}{1} + \frac{1}{4} + \frac{1}{8} \right]$$

$$E_0 = (-10 \text{ V}) \left[\frac{1}{1.375} \right] = -7.27 \text{ volts}$$

17.5 R-2R RESISTANCE LADDER CIRCUITS

Although not revealed by the idealized equations, the binary weighted resistance ladder suffers from a serious drawback in actual practice. The values of the input resistors tend to become very large and very small at the ends of the range as the bit length of the input word becomes longer. If R is set to 10 kΩ (a popular value), then $R8$ will be 1.28 megohms. If we assume a reference potential E of 10.00 volts dc, then $I8$ will be only 7.8 *microamperes*. Most common nonpremium-grade operational amplifiers will not be able to resolve signals that low from the inherent noise. As a result, the bit length of the binary weighted ladder is severely limited. Few of these types of converters are found with more than 6- or 8-bit word lengths.

In commercial DACs, all of the resistors have a value of either R or

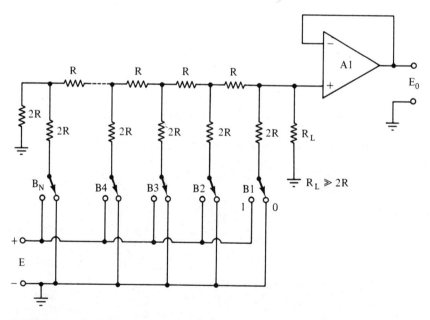

FIGURE 17-2. R-2R resistor ladder DAC circuit.

Section 17.5 / R-2R RESISTANCE LADDER CIRCUITS

$2R$. The gain of the amplifier is unity, so E_0 can be expressed as

$$E_0 = (E) \sum_{i=1}^{n} \frac{a_i}{2^i} \qquad (17\text{-}7)$$

(provided that $R_L \gg R$, so that the voltage divider effect between the ladder and R_L can safely be neglected).

Example 17-2

A four-bit DAC using the R-$2R$ technique has a 5.00-volt dc reference potential. Calculate E_0 for the input word 1011_2.

Solution

$$E_0 = E \sum_{i=1}^{n} \frac{a_i}{2^i} \qquad (17\text{-}7)$$

$$E_0 = (5 \text{ V}) \left[\frac{1}{2^1} + \frac{0}{2^2} + \frac{1}{2^3} + \frac{1}{2^4} \right]$$

$$E_0 = (5 \text{ V}) \left[\frac{1}{2} + 0 + \frac{1}{8} + \frac{1}{16} \right]$$

$$E_0 = (5 \text{ V})(0.688) = \mathbf{3.44 \text{ volts}}$$

The full-scale output voltage for any DAC using the R-$2R$ resistor ladder is given by

$$E_{fs} = \frac{E(2^n - 1)}{2^n} \qquad (17\text{-}8)$$

where E_{fs} = the full-scale output potential in volts (V)
E = the reference potential in volts (V)
n = the bit length of the digital input word

Example 17-3

Find the full-scale output potential for an eight-bit DAC with a reference potential of 10.00 volts dc.

Solution

$$E_{fs} = \frac{E(2^n - 1)}{2^n} \qquad (17\text{-}8)$$

$$E_{fs} = \frac{(10\ V)(2^8 - 1)}{2^8}$$

$$E_{fs} = \frac{(10\ V)(2^7)}{(2^8)}$$

$$E_{fs} = \frac{(10\ V)(255)}{(256)} = \mathbf{9.96\ volts}$$

The output of a DAC cannot change in a continuous manner, because the input is a digital word; i.e., it can exist only in certain discrete states. Each successive binary number changes the output an amount equal to the change created by the least significant bit (LSB), which is expressed by

$$\Delta E_0 = \frac{E}{2^n} \qquad (17\text{-}8)$$

So, for the DAC in Example 17-3, E_0 would be

$$\Delta E_0 = \frac{(10\ V)}{2^8}$$

$$\Delta E_0 = \frac{(10\ V)}{(256)} = \mathbf{40\ mV}$$

ΔE_0 is often called the 1 LSB value of E_0, and is the smallest change in output voltage that can occur. It is interesting that, if we let 0 volts represent 00000000_2 in our eight-bit system, then the maximum value of E_0 at 11111111_2 will be 1 LSB less than E (confirmed by the result of Example 17-3).

There are numerous commercial DACs on the market in IC, function module block, and equipment form. The reader should consult manufacturer's catalogues for appropriate types in any given application.

17.6 SUMMARY

1. Analog-to-digital converters (ADC) convert analog voltages or currents to binary words.

2. Digital-to-analog converters (DAC) convert binary words to proportional voltages and currents.
3. There are two common forms of DAC, distinguished by the type of resistance ladder used: binary weighted or R-$2R$.

17.7 RECAPITULATION

Now go back and try answering the questions at the beginning of the chapter. When you are finished, answer the questions and work the problems given below. Place a mark beside each problem or question that you cannot answer, and then go back to the text and reread appropriate sections.

17.8 QUESTIONS

1. Define in your own words the purposes of the DAC.
2. Draw a circuit for a binary weighted resistor ladder DAC.
3. Draw a circuit for an R-$2R$ resistor ladder DAC.
4. The output of a DAC is a _____ function of voltage or current.

17.9 PROBLEMS

1. An eight-bit DAC using a binary weighted resistor ladder has a +7.5-volt dc reference potential. Find the output potential for the input word 10110111_2 if $R_f = R$.
2. An eight-bit DAC using an R-$2R$ ladder has a +2.56-volt dc reference potential. Find the full-scale output potential.
3. Find the 1 LSB output voltage for the DAC in Problem 2.

17.10 REFERENCES

1. Joseph J. Carr, *Elements of Electronic Instrumentation & Measurement*. Reston, Va.: Reston Publishing Co. 1979.
2. Joseph J. Carr, *Microprocessor Interfacing Handbook: A/D-D/A*. Blue Ridge Summit, Pa.: TAB Books, 1980.
3. D.H. Sheingold, *Analog-to-Digital Conversion Handbook*. Norwood, Mass.: Analog Devices, Inc., 1979.

18
Basics of Data Conversion—A/D

18.1 OBJECTIVES

1. To list the principal types of analog-to-digital converter (A/D).
2. To understand the applications and limitations of the various types of A/D converter.
3. To describe the operation of A/D converters.
4. To list advantages and disadvantages of the various A/D converters.

18.2 SELF-EVALUATION QUESTIONS

Before studying the material in this chapter, try answering the questions given below. These questions test your knowledge of the subject. If you cannot answer any particular question, then place a check mark beside it and look for the answer as your read the text.

1. Describe in your own words the operation of a *servo* A/D converter.
2. An eight-bit A/D converter has a maximum input voltage rating of 2.55 volts. What binary word will represent this potential if the A/D converter is wired for unipolar positive operation?
3. What is the common name for an A/D converter that uses an *integrator* stage? What is the most common use of this type of A/D converter?
4. Which will convert a fullscale potential the fastest, servo or successive approximation?

18.3 INTRODUCTION TO ANALOG-TO-DIGITAL CONVERSION

The purpose of the analog-to-digital converter is to produce an output binary word that is proportional to an analog input signal. The output of

Section 18.5 / SERVO ADC CIRCUITS

the A/D converter is input to a computer, where it is used in the various processes and calculations made by the computer. The type of A/D converter selected often depends upon the job that must be done, so sufficient concern must be shown for the specifications.

18.4 TYPES OF A/D CONVERTERS

There are several different *basic* analog-to-digital converter circuits, and they all have their own typical applications. Although the circuitry varies from one model to another, it is often the *speed* that is the most important specification in some applications, cost being second in importance. The *dual-slope integrator* is often very low in cost but is very slow. Similarly, the *flash* converter is very rapid (video rates), but it is difficult to decode into a format that the computer likes, and must often be used with RF-like connecting circuitry. The *successive approximation* converter usually acquits itself nicely, except that it can be too complex and costly. It will, however, easily out perform most versions of the other types of converter.

In this chapter, we will consider these basic converter types: *servo* (also called *binary ramp*), *successive approximation, flash* (or *parallel*), *dual-slope integration,* and *voltage-to-frequency converters.*

18.5 SERVO ADC CIRCUITS

The *servo* ADC circuit (also called *binary counter* or *ramp* ADC circuit) uses a binary counter to drive the digital inputs of a DAC. A voltage comparator keeps the clock gate to the counter open as long as $E_0 \neq E_{in}$.

An example of such a circuit is the eight-bit ADC in Figure 18-1(a), while the relationship of E_0 and E_{in} relative to time is shown in Figure 18-1(b).

Two things happen when a *start* pulse is received by the control logic circuits: the binary counter is reset to 00000000_2, and the gate is opened to allow clock pulses into the counter. This will permit the counter to begin incrementing, thereby causing the DAC output voltage E_0 to begin rising (Figure 18-1(b).) E_0 will continue to rise until $E_0 = E_{in}$. When this condition is met, the output of the comparator drops *low,* turning off the gate. The binary number appearing on the counter output at this time is proportional to E_{in}.

The control logic section senses the change in comparator output level, and uses it to issue an *end-of-conversion* (EOC) pulse. This EOC pulse is used by instruments or circuitry connected to the ADC to verify that the output data are valid.

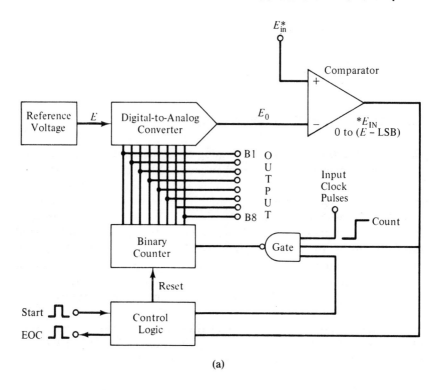

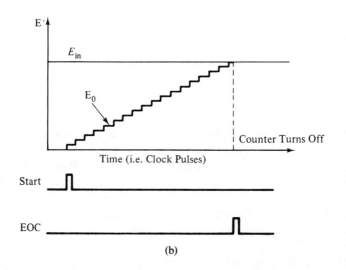

FIGURE 18-1. (a) Servo ADC circuit. (b) Operation of the servo-type ADC circuit.

Section 18.6 / SUCCESSIVE APPROXIMATION ADC CIRCUITS

The *conversion time* T_c of an ADC such as this depends upon the value of E_{in}, so when E_{in} is maximum (i.e., full scale), so is T_c. Conversion time for this type of ADC is on the order of 2^n clock pulses for a full-scale conversion.

18.6 SUCCESSIVE APPROXIMATION ADC CIRCUITS

The conversion time of the servo ADC is too long for some applications. The successive approximation (SA) ADC is much faster for the same clock speed; i.e., it takes $(n + 1)$ clock pulses instead of 2^n. For the eight-bit ADC that has been our example, the SA type of ADC is 28 times *faster* than the *servo* ADC.

The basic concept of the SA ADC circuit can be represented by a platform balance, such as Figure 18-2, in which a full-scale weight W will deflect the pointer all the way to the left when pan 2 is empty.

Our calibrated weight set consists of many separate pieces, which weigh $W/2$, $W/4$, $W/8$, $W/16$, etc. When an unknown weight W_x is placed on pan 2, the scale will deflect to the right. To make our measurement, we start with $W/2$, and place it on pan 1. Three conditions are now possible:

$W/2 = W_x$ (scale is at zero)

$W/2 > W_x$ (scale is to the left of zero)

$W/2 < W_x$ (scale is to the right of zero)

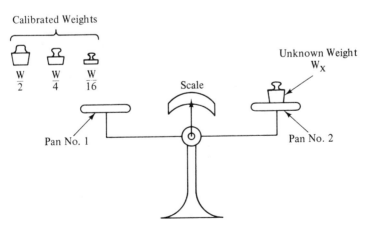

FIGURE 18-2. Successive approximation ADC circuits are like the platform balance.

If $W/2 = W_x$, then the measurement is finished, and no additional trials are necessary. But if $W/2$ is less than W_x, then we must add more weights in succession ($W/4$, then $W/8$, etc.) until we find a combination equal to W_x.

If, on the other hand, $W/2$ is greater than W_x, then we must *remove* the $W/2$ weight, and in the second trial start again with $W/4$. This procedure will continue until a combination equal to W_x is found.

In the SA ADC circuit we do not use a scale, but a shift register, as in Figure 18-3. A successive approximation register (SAR) contains the control logic, a shift register, and a set of output latches, one for each register section. The outputs of the latches drive a DAC.

A *start* pulse sets the first bit of the shift register *high*, so the DAC will see the word 10000000_2, and therefore produces an output voltage equal to one-half of the fullscale output voltage. If the input voltage is greater than $\frac{1}{2}E_{fs}$, then the $B1$ latch is set *high*. On the next clock pulse, register $B2$ is set high for trial 2. The output of the DAC is now ¾-scale. If, on any trial, it is found that $E_{in} < E_0$, then that bit is reset *low*.

Let us follow a three-bit SAR through a sample conversion. In our example, let us say that the full-scale potential is 1 volt, and E_{in} is 0.625 volt. Consider Figure 18-4.

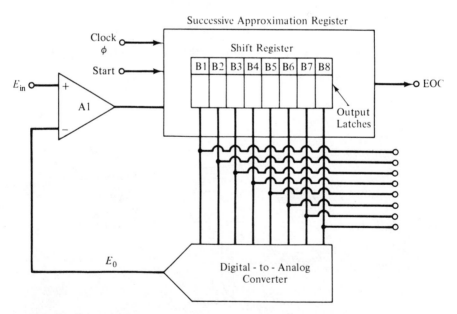

FIGURE 18-3. Successive approximation ADC circuit.

Section 18.6 / SUCCESSIVE APPROXIMATION ADC CIRCUITS

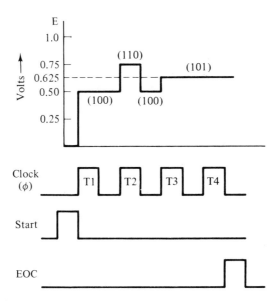

FIGURE 18-4. Timing diagram for Figure 18-3.

Time t_1: The *start* is received, so register $B1$ goes *high*. Output word is now 100_2, so $E_0 = 0.5$ V. Since E_0 is less than E_{in}, latch $B1$ is set to "1," so at the end of the trial, the output word remains 100_2.

Time t_2: On this trial (which starts upon receiving the next clock pulse), register $B2$ is set *high*, so the output word is 110_2. Voltage E_0 is now 0.75 volt. Since E_{in} is less than E_0, the $B2$ latch is set to "0," and the output word reverts to 100_2.

Time t_3: Register $B3$ is set *high*, making the output word 101_2. The value of E_0 is now 0.625 volt, so $E_{in} = E_0$. The $B3$ register is latched to 1, and the output word remains 101_2.

Time t_4: Overflow occurs, telling the control logic to issue an EOC pulse. In some cases the overflow pulse *is* the EOC pulse.

Note that in the example, we had a three-bit SAR, so by our $(n + 1)$ rule, required four clock pulses to complete the conversion. The SA type of ADC was once regarded as difficult to design because of the logic required. But today IC and function blocks are available that use this technique, so the design job is reduced considerably. The SA technique can be implemented in software under computer control using only an external DAC and comparator. All register functions are handled in the software (program).

18.7 PARALLEL CONVERTERS

The parallel ADC circuit (Figure 18-5) is probably the fastest type of ADC known. In fact, some texts call it the "flash" converter in testimony to its speed. It consists of a bank of $(2^n - 1)$ voltage comparators biased by reference potential E through a resistor network that keeps the individual comparators 1 LSB apart. Since the input voltage is applied to all of the comparators simultaneously, the speed of conversion is essentially the

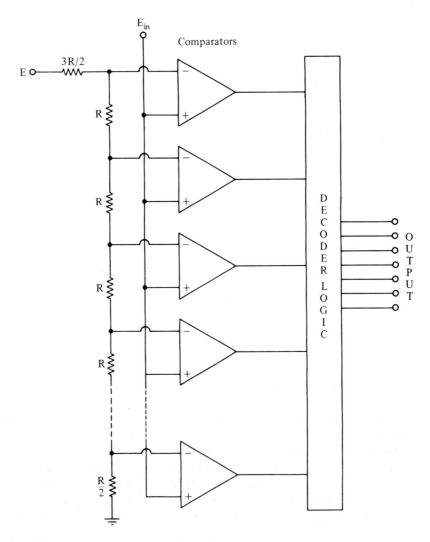

FIGURE 18-5. Parallel or "flash" ADC circuit.

Section 18.8 / VOLTAGE-TO-FREQUENCY CONVERTERS

slewing speed of the *slowest* comparator in the bank, and the decoder propagation time (if logic is used). The decoder converts the output code to binary code, or possibly BCD in some cases.

18.8 VOLTAGE-TO-FREQUENCY CONVERTERS

A voltage-to-frequency (V/F) converter is a voltage-controlled oscillator (VCO) in which an input voltage E_{in} is represented by an output frequency F. An ADC using the V/F converter is shown in Figure 18-6. It consists of no more than the VCO and a frequency counter. The display, or output states, on the counter gives us the value of E_{in}.

Voltage-to-frequency converters are used mainly where economics dictate serial transmission of data from a remote collection point to the instrument. Such data can be transmitted by wire or radio communications channels. Another application is the tape recording of analog data that is, in itself, too low in frequency to be recorded.

The inverse procedure, F/V conversion, is a form of DAC, in which an input *frequency* is converted to an output *voltage*.

The type of V/F converter shown in Figure 18-6 is a simplistic circuit and generally too crude for practical application. Most commercial V/F converters are variations on the circuit shown in Figure 18-8, which is a version of a product made by Teledyne/Philbrick.

The V/F converter is basically a freerunning astable multivibrator in which the operating frequency is controlled by the input voltage. This circuit may also be called a *voltage controlled oscillator* (VCO), although in common application the VCO has too limited a range for serious A/D work.

The input operational amplifier (A1) is connected as a driver for the precision current source, transistor Q1. The collector current of this NPN transistor is a linear function of the input currents applied to amplifier A1.

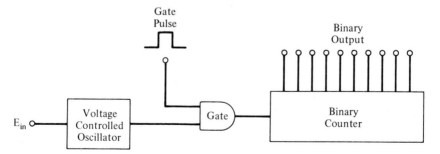

FIGURE 18-6. Voltage-to-frequency converter ADC.

The collector current (I) is used to charge timing capacitor C. We may conclude, therefore, that the charging rate of capacitor C is also a linear function of the input currents applied to $A1$. Note that a capacitor that is charged from a *constant current source* (e.g., Q1) will have a linear ramp for the charging voltage, and *not* the exponential ramp expected for *constant voltage source* circuits that are normally considered in elementary texts.

Schmitt triggers are special cases of comparator circuits in which the output snaps HIGH when the analog input voltage crosses a certain threshold in the positive-going direction, and then drops LOW again when the input voltage falls below another threshold in the negative-going direction. The capacitor voltage, E_c, is used as the analog input to the Schmitt trigger circuit in Figure 18-7(a). When the capacitor voltage reaches the trip threshold ($E2$ in Figure 18-7(b)), the Schmitt trigger will go HIGH, and (a) trigger the monostable multivibrator (one-shot) that produces the V/F converter output pulse, and (b) activate the *precision charge dispenser* (PCD) that will discharge capacitor C to the same level (non-zero) each time it is turned on.

The timing waveforms for the V/F converter are shown in Figure 18-7(b). The voltage across the capacitor will rise from its quiescent, non-zero, value until it reaches E2. This voltage is the Schmitt trigger threshold, so it will trigger the circuit monostable multivibrator into producing an output pulse. At this same time, the PCD is also activated, causing the capacitor to discharge to the quiescent value, $E1$.

The slope of the capacitor charging waveform, hence the frequency of the monostable multivibrator, is a function of the collector current of transistor Q1, current I. This current is, in turn, a function of the analog input voltages (or currents, depending upon which input terminal is used) to amplifier A1, we can see that the output frequency is proportional to the input voltage. Typical V/F converters are commonly available with frequencies up to 10 kHz, with premium models or grades with 100 kHz frequencies. Some special units are known that will output frequencies up to 1000 kHz.

18.9 DUAL-SLOPE INTEGRATION

The block diagram of a dual-slope integrator is shown in Figure 18-8(a), while associated waveforms are shown in Figure 18-8(b).

The heart of the circuit is the operational amplifier *integrator* (see Chapter 12) consisting of operational amplifier $A1$, plus $R1$ and $C1$, plus a *voltage comparator* ($A2$). The output of the comparator will remain LOW

Section 18.9 / DUAL-SLOPE INTEGRATION

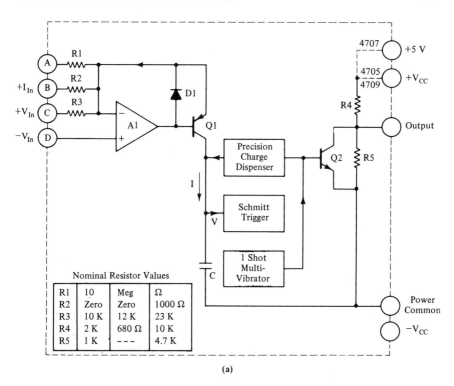

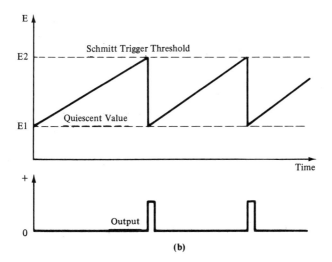

FIGURE 18-7. Voltage-to-frequency converter. (a) Circuit. (b) Waveforms.

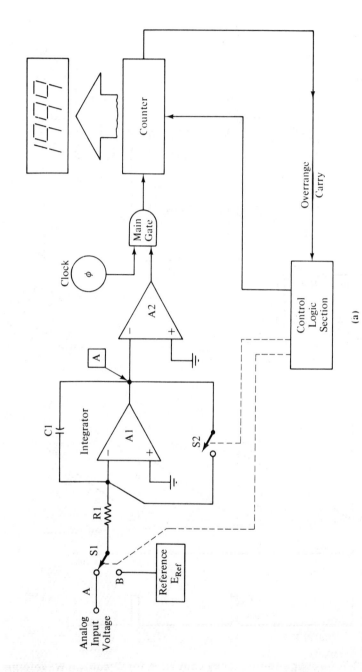

FIGURE 18-8. (a) Dual-slope integrator for electronic voltmeters.

Section 18.9 / DUAL-SLOPE INTEGRATION

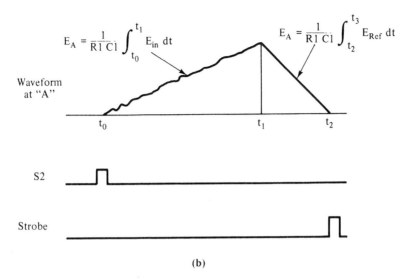

(b)

FIGURE 18-8. (b) Timing of the dual-slope integrator.

if the integrator output is zero, and HIGH if the integrator output is more than a few millivolts above ground potential.

At the beginning of the conversion cycle, the control logic section momentarily closes electronic switch $S2$ so that the charge on capacitor $C1$ goes to zero, and also ensures that $S1$ is set to position "A."

If $S1$ is in position "A," the integrator input is connected to the input voltage source, causing the voltage at the integrator output to begin rising [see time t_0 in Figure 18-8(b)]. As soon as E_A rises a few millivolts the comparator output snaps HIGH, enabling the gate to pass clock pulses to the digital counter (Chapter 6) section. The counter is allowed to overflow [$t1$ in Figure 18-8(b)], and the output *carry* pulse from the counter is used to tell the control logic section to switch $S1$ to position "B." This action connects the integrator input to a *precision reference* voltage source. The polarity of the input current created by the reference is such that it begins to *discharge* the integrator capacitor at a *constant* rate. The counter, meanwhile, has continued to increment, passing through 0000 at time $t1$, and continuing to accumulate clock pulses until E_A is back down to zero.

The value of E_A at time t_4 was proportional to the value of E_{in}. At the same instant the count was 0000. Since the counter continues to increment as the integrator discharges (t_1 to t_2), the *count* at the instant the gate is closed (when $E_A = 0$) is also proportional to E_{in}. By correct scaling the count will be numerically the same as the potential applied to the input. To recapitulate:

1. At time t_0 switch $S2$ is closed briefly to dump any residual charge in $C1$, and $S1$ is set to position "A."
2. The integrator begins to charge due to current $E_{in}/R1$, so E_A begins rising from zero.
3. As soon as E_A is greater than zero, the output of the comparator goes HIGH, enabling the main gate to pass clock pulses into the counter.
4. The counter increments until it overflows at time t_1, and the overflow pulse causes $S1$ to switch to position "B," applying the reference voltage to the input of the integrator. At this instant the count is 0000.
5. From times t_1 to t_2 the integrator *discharges* under the influence of current $-E_{ref}/R1$; meanwhile, the counter continues to increment.
6. At time t_2 the comparator shuts off the flow of clock pulses through the gate. The count accumulated between t_1 and t_2 represents the input voltage.
7. The control logic section issues a *strobe* pulse at time t_2 to update the counter's digital display.

The dual-slope integrator is one of the slower types of ADC circuit, but offers several advantages when used in DVM circuit including relative immunity to noise riding on the input voltage and relative immunity to error due to inaccuracy or long term drift in the clock frequency. The 10 to 40 ms conversion cycle time poses no disadvantage in most DVM applications.

18.10 WHAT DOES THE DUAL-SLOPE DVM MEASURE?

A dc digital voltmeter tends to read the *average* input voltage over the period of integration. This feature makes the DVM particularly useful for measuring noisy signals. It also tends to average the pulsations from a rectifier-type ac-to-dc converter, provided that $t_2 - t_0$ is greater than about 18 ms.

The accuracy of a digital voltmeter, or DMM based on a DVM, can be quite good. In fact, it can exceed the accuracy of analog multimeters. You will see terms such as "2½-digit," "3½-digit," and "4½-digit" in the advertisements for DVM/DMM products, and will wonder what is meant by a "½" digit. This term refers to the fact that the *most significant* digit can only be a "0" or a "1," while all other digits can be anything between "0" and "9." Such terminology indicates that the meter can read 100 percent overrange from its basic range. For example, a 3½-digit DVM will read 0–1999 mV, while its basic range is only 0–999 mV. If this range is exceeded, then the "1" lights up; otherwise it remains darkened.

18.11 SUMMARY

1. Analog-to-digital converters produce a binary output that is proportional to an analog input voltage or current.
2. Several different types of A/D converter are in common use: servo ramp, successive approximation, dual-slope integrators, flash or parallel, and V/F converters.
3. The resolution of the A/D converter is proportional to the bit-length of the output word.
4. Feedback A/D converters are those that use a DAC circuit in the conversion process.

18.12 RECAPITULATION

Now go back and try answering the questions at the beginning of the chapter. When you are finished, answer the questions and work the problems given below. Place a mark beside each problem or question that you cannot answer, and then go back to the text and reread appropriate sections.

18.13 QUESTIONS

1. Define in your own words the purposes of an analog-to-digital converter.
2. Draw a block diagram for a V/F converter using a PCD element.
3. List two types of A/D converter in which a DAC is a critical element.
4. Describe in your own words the operation of a binary ramp (servo) type A/D converter.
5. Describe in your own words the operation of a successive approximation A/D converter.
6. Draw the block diagram for a servo A/D converter.
7. Draw a block diagram for a dual-slope integrator.
8. Draw a block diagram for a successive approximation A/D converter.
9. A very fast type of A/D converter that uses a bank of voltage comparators is called the _____ or _____ A/D converter.
10. Feedback A/D converters are those that use a _____ in the conversion process.
11. In a continuously cycling A/D converter the end-of-conversion (EOC) pulse from the last conversion cycle becomes the _____ pulse for the next conversion cycle.

18.14 PROBLEMS

1. Find the conversion cycle time for a fullscale input potential in the eight-bit servo A/D converter if a 1 mHz system clock is used.
2. Find the conversion cycle time for a fullscale input potential in the eight-bit successive approximation A/D converter if a 1 mHz clock is used.
3. Find the conversion cycle time for a fullscale potential in a 10-bit binary ramp A/D converter when a 2.5 mHz clock is used.
4. An eight-bit A/D converter in unipolar positive operation represents the fullscale input potential of 2.55 volts with the word hFF. Find the binary and hexadecimal (h) representations for (a) half-scale potential, (b) zero potential, and (c) 3/5-scale potential.
5. The A/D converter in problem 4 sees an input potential of 1.70 volts. What is the hexadecimal and binary code that will represent this potential?
6. The A/D converter in problem 4 has an output code of h0F. What is the input voltage?

18.15 REFERENCES

1. Joseph J. Carr, *Elements of Electronic Instrumentation and Measurement*. Reston, Va.: Reston Publishing Co., 1979.
2. Joseph J. Carr, *Microprocessor Interfacing Handbook: A/D-D/A*. Blue Ridge Summit, Pa.: TAB Books, 1980.
3. D.H. Sheingold, *Analog-to-Digital Conversion Handbook*. Norwood, Mass.: Analog Devices, Inc.

19
Data Acquisition Systems

19.1 OBJECTIVES

1. To learn the different types of data acquisition system.
2. To understand some of the advantages and disadvantages of each.
3. To learn how to select a system.

19.2 SELF-EVALUATION QUESTIONS

Before studying the material in this chapter, try answering the questions given below. These questions test your knowledge of the subject matter. If you cannot answer a particular question, then place a check mark beside it and look for the answer as you read the text.

1. List several different types of product that qualify for the title data acquisition system.
2. What factors affect the selection of the data acquisition system?
3. What is the purpose of a multiplexer in the data acquisition system?
4. What is the cause of "walk" or "data creep" in a data acquisition system?
5. What is the purpose of the *Sample & Hold* circuit in a data acquisition system?

19.3 INTRODUCTION

What is a *data acquisition system?* Obviously, it is some kind of system that is used to acquire data. The question is not really trivial because there are a number of products on the market that qualify for the designation "data acquisition system." One kind of DAS is a rack-mounted device intended to accompany a minicomputer or large microcomputer in a 19-inch instrument rack. In another case, the DAS is a printed circuit board that plugs into a microcomputer or minicomputer and serves to convert analog voltage or current levels into binary words. Still another DAS may be a hybrid function module that is complete within itself and needs only mounting on a printed board or in a cabinet to function with a microcomputer system. A last category of DAS is an integrated circuit that contains almost all of the components needed to make it work but may still require some external circuitry.

In general, however, we can define the data acquisition system as a device or circuit that converts analog data into digital format and is ready for direct interfacing with a micro/mini computer. The DAS may also contain DACs, as well as ADCs, so that an analog signal may be created from a digital output on the computer. That DAC feature, however, is an option that is not totally necessary to our definition. The DAS may be a single channel affair or it may be multi-channel. Eight and sixteen channel DASs are readily available, and some manufacturers offer DAS in up to 64 channels. Such a DAS will convert up to 64 independent channels of analog data into binary format and input it into the computer.

Figure 19-1 shows two typical printed circuit board data acquisition systems. These PC boards are made by Burr-Brown and Motorola, Inc., respectively, and are intended to be plugged into standard computers.

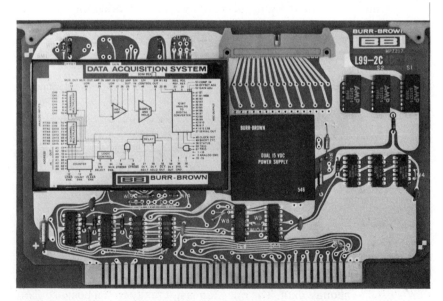

FIGURE 19-1(a). Data acquisitions system. (*Courtesy of Burr-Brown.*)

Section 19.4 / COMPONENTS OF THE DATA ACQUISITION SYSTEM

FIGURE 19-1(b). Data acquisitions system. (*Courtesy of Motorola, Inc.*)

These companies and others manufacture several different plug-compatible DAS PCBs that will plug directly into certain standard microcomputer frames.

19.4 COMPONENTS OF THE DATA ACQUISITION SYSTEM

We are still stuck with our original question: what is a data acquisition system? The DAS is a device or circuit that will contain an analog-to-digital converter (complete), plus instrumentation amplifiers (as needed), *multiplexing* circuits to allow multichannel operation, interfacing circuits, and the control logic section.

Perhaps we can home in on the definition of the DAS by considering the block diagrams for a couple of examples. Figure 19-2 shows a 16-channel, 8-bit DAS made using monolithic integrated circuit techniques: the *Datel* DAS-952R. There is only one A/D converter in the DAS-952R, so it must be switched between the 16 different analog input channels. The *analog multiplexer* is the section that accomplishes channel selection. The *channel address control* and the *inhibit control/address enable* terminal determines the active channel. A LOW on the inhibit control turns off all channels, while a HIGH will cause the channels to be addressed. In a microcomputer DAS, then, we can connect the inhibit pin to

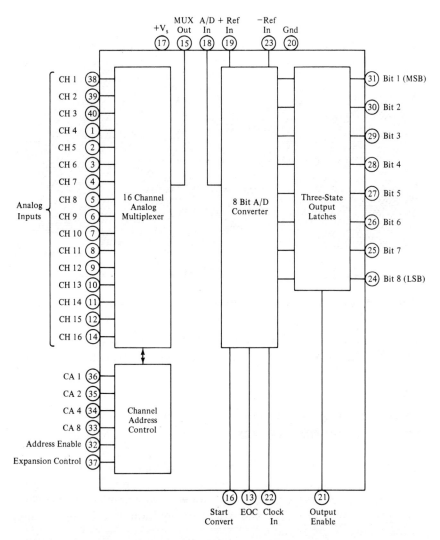

FIGURE 19-2. Chip-form data acquisition system. (*Courtesy of Datel.*)

an addres decoder or similar device select circuit, allowing the DAS to be turned on and off all at once. The channel address pins (CA1, CA2, CA4, and CA8) select the analog channels 1 through 15 according to a straight binary code in which 0000 selects channel no. 1, and 1111 selects channel no. 16.

In this particular DAS, the output of the multiplexer and the input of the eight-bit A/D converter are brought to separate pins on the device

Section 19.4 / COMPONENTS OF THE DATA ACQUISITION SYSTEM

package. These pins are normally joined together, but sometimes it is wise to use a Sample & Hold circuit between the two sections.

The A/D converter has the usual complement of control pins, namely a system clock (which could be the computer clock), a *start convert* terminal, and an *end of conversion* (EOC) signal. When a pulse is applied to the start terminal, the A/D will begin its process. At the end of the conversion period an EOC pulse will appear at the EOC terminal. This pulse would be used by the computer to indicate when the data is ready for input to the computer.

The Datel DAS-952R system can be interfaced directly with the microcomputer data bus because it has a three-state (i.e., tri-state) output latch bank. These latches will go to high impedance when the *output enable* is LOW. In this condition, the DAS outputs will float harmlessly across the data bus without interfering with other signals.

Another example is the Datel HDAS-8 and HDAS-16 shown in Figure 19-3. This DAS is a 12-bit system, and can be configured in either 16 or eight channel versions. It will provide either 16 single-ended channels or eight differential channels. The data acquisition time for the built-in A/D converter (plus amplifier settling time and sample & hold aperture time) is on the order of 20 microseconds, which permits a data throughput of 50 kHz. The output coding of the A/D converter can be either straight binary for unipolar operation or offset binary for bipolar operation. Note that the tri-state output latches are grouped in a 3X4 configuration, i.e., three banks of four bits each. This configuration allows us to connect the 12-bit DAS to an eight-bit microcomputer. The data is then output in at least two different operations. For example, we could connect bits B1 through B8 to the B1 through B8 lines of the data bus. The enable lines for B1-B4 and B5-B8 are then connected together so that these bits will work as a single eight-bit word. We can then also connect outputs B9 through B12 to data bus lines B1 through B4. This means that B1 of the DAS is in parallel with B9 and both are connected to B1 of the computer data bus. Similarly, B2/B10, B3/B11, and B4/B12 are connected to data bus lines B2, B3, and B4, respectively. A second read operation, sequential to that which input the lower order eight bits, will allow us to input the high order bits. Microcomputers that have instructions and register pairs that assist double precision (two byte) data words are especially usable in this situation. Chapter 20 will deal further with the problem of interfacing greater-than-8-bit data converters to an eight-bit bus.

The HDAS-8/16 data acquisition systems contain a programmable gain instrumentation amplifier. A single external resistor will make the gain of this differential amplifier vary from 1 to 1000. This particular feature makes the HDAS-8/16 particularly useful for a wide range of applications because it will accommodate a wide range of signals. In some cases,

DATA ACQUISITION SYSTEMS / Chapter 19

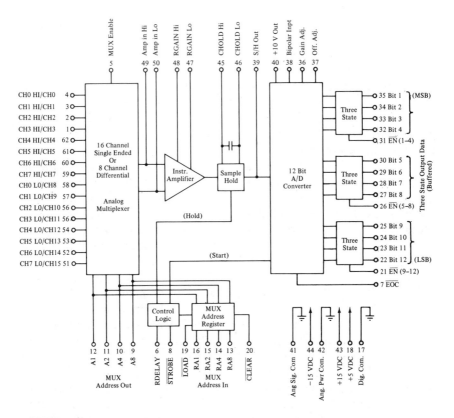

FIGURE 19-3. Data acquisition system. (*Courtesy of Datel.*)

the DAS will interface its input with the output of some other instrument or system. Such a system might have a 0 to 1 volt, or 0 to 10 volt, output voltage range. In many research laboratories and industrial plant control systems this situation is found due to the existence of a lot of other instruments. The job of the microcomputer interface engineer is to select the correct DAS to accommodate all of those pre-existing systems. In still other cases, the DAS may have to interface directly with a thermocouple, strain gage, or other transducer. The output potentials of these transducers are typically microvolts or millivolts. As a result, an instrumentation amplifier with a high gain is needed. A 10 mV signal into a gain-of-1000 amplifier will yield a range of 0 to 10 volts.

The Sample & Hold is used to keep the analog signal constant while the A/D converter is making the conversion. This feature is especially needed on the successive approximation form of data converter.

Some systems will use a series of Sample & Hold circuits, one each for the analog channels. This can be done when the conversion time to make all 16 channels into digital form is too long for the rate of change of the various data signals. This is especially necessary in some scientific experiments, such as might be found in a medical laboratory. Many of the parameters are interrelated, and in fact it is these correlations that the system user is seeking. In order to make sure that correlated data stays correlated, it is necessary to sample all lines at the same time and then make the various A/D conversions. This procedure allows us to have related analog data signals essentially converted at the same time. But this is not for free. In some cases, especially where a slow A/D converter is used for the number of analog channels, we see a phenomenon called *data creep* or *data walk*. It occurs because of the droop in the S&H circuits. As discussed in Chapter 14, a problem with all capacitor-stored S&H circuits is the slow discharge of the capacitor during the store mode. If this droop is sufficiently large, then we will find that the data at the last conversion in a multi-channel DAS is not as valid as the first conversion. We can test for "creep" by connecting all inputs to the same signal source, and then sampling. If there is no creep, then the A/D conversion process will yield the same binary value for all parallel channels ($\pm$1-LSB because of normal A/D uncertainties).

19.5 DAS SELECTION

It is all too easy to overspecify data acquisition systems. The ads for these devices tout such features as conversion time, throughput, number of channels, and other factors. But over-speced DASs cost a lot of money and could easily put an otherwise viable project into the red. On the other hand underspecified DASs are also money wasters because they will not do the job. Obviously, we must have some means for zeroing in on the specs that are most pertinent to the job.

Of course, the first job is to determine the form that the DAS will take. Are we going to need a rack-mounted DAS that can be mounted with a mini or microcomputer cabinet? Such a system would often be cheaper in the long run than building an equivalent data acquisition system from one of the other products. In other cases, however, we will want to buy a pre-made printed circuit board (see again Figure 19-1) that is designed to plug into the microcomputer that we have selected (or was selected for us). In still other cases, we will want to buy either an integrated circuit or hybrid function module DAS and mount it on a PC board ourselves. This is the feature that would be followed if we had a bastard computer or one that is not popular enough to have attracted the attention of the PC board DAS makers. We would also use this system if the DAS was to be remote or part of another instrument or system, and then only the data would be transmitted to the computer.

We would also have to consider the interfacing for the DAS. In cases where the DAS is mounted inside of the computer, plugged into the computer bus, then it is probably best to memory map the DAS. In this system, the eight to 16 channels are treated as memory locations. In some versions, only the lowest channel need be specified because it uses a memory address boundary system. The DAS is located exactly at some 4K boundary in the computer memory space. Channel no. 1 has the address of the boundary selected, while the remaining 15 channels are automatically placed in the next 15 memory slots above the boundary. For example, in a certain B-B product, there are programming pins that are used to specify the starting address for channel no. 1. These are programmed by hardwiring either HIGH or LOW the correct pins for the address selected.

The second interfacing method, used where memory-mapping is not possible, is to connect the DAS to a series of I/O ports. This would be done in cases where the memory is full or the DAS is remote. It is not usually too good an idea to extend the data and address buses of the microcomputer indefinitely.

Thus far, we have determined two factors that must be selected by the engineer specifying the DAS: form of the DAS and interfacing method.

Next, we must consider the input voltage range and the output coding. If the input voltage range is excessive, then either another DAS must be selected or an attenuator provided between the signal source and the DAS input channel. If, on the other hand, the signal level is too low for efficient use of the admissible codes, then we must provide an amplifier for that channel. In some cases, it will be necessary to attenuate some channels, amplify others, and leave the rest alone. This necessity points up the utility of DASs that contain built-in programmable-gain instrumentation amplifiers. When a channel is selected, then the proper gain for that channel is also cranked into the DAS.

Select the output code according to the needs of the system. For example, we can use ordinary straight binary for unipolar systems. But if the analog signal is bipolar, then we must either apply an offset voltage that makes it unipolar or provide the coding scheme that will account for the bipolar input voltage. Remember, there is more than one system for representing bipolar data, and the one selected depends in part on how dearly we need to know when the value is exactly zero.

The conversion time, throughput rate, and sampling rate are all interrelated parameters. We know that there are certain constraints placed

Section 19.5 / DAS SELECTION

upon the sampling rate. A well known theorem tells us that the sample rate must be at least twice the highest frequency component in the waveform. For example, if the Fourier series for the waveform has significant "harmonics" to 500 Hz, then a 1000 samples/second sampling rate is needed.

Throughput rate and sampling rate can cost a lot of money. If, for example, we have a lot of 10 kHz waveforms, then it will be necessary to have some very fast DASs. But if most of the parameters in the system have much lower frequency components, and only one or two waveforms have a high frequency content, it might be wise to use a low-rate DAS, and then buy a couple of faster A/D converters to serve only the high frequency channels. Such a procedure would usually be cheaper than specifying all 16 (or more) channels to have the same frequency as needed for the single high frequency channel.

20
Data Converter Interfacing

20.1 OBJECTIVES

1. To learn methods for connecting digital-to-analog converters (DAC) to microcomputers and microprocessors.
2. To learn methods for connecting DACs to microcomputers with fewer output port bits than the DAC.
3. To connect analog-to-digital converters to microcomputers and microprocessors.
4. To connect and utilize the A/D converter service signals in microprocessor-based instrumentation.

20.2 SELF-EVALUATION QUESTIONS

Before studying the material in this chapter, try answering the questions given below. These questions test your knowledge of the subject matter. If you cannot answer a particular question, then place a check mark beside it and look for the answer as you read the text.

1. Draw a simple schematic for connecting a DAC to the output port of a microcomputer.
2. Define the typical "service signals" used on A/D converters.
3. How may an A/D converter be connected to make continuous conversions?
4. The _____ _____ places a fundamental limitation on the number of samples per second that may be realized with any given data converter.

20.3 DATA CONVERTERS AND DATA CONVERSION

Computer enthusiasts tend to believe that all jobs are best done on digital instruments . . . and to some extent this may be true. But it is a fact of life that most of nature is *analog*. The vast majority of transducers, as well as naturally "transducable" events, produce analog voltage or current signals as outputs. When, for example, we want to measure a fluid pressure (medical instruments, factory process instruments, etc.), we will find a number of different appropriate transducers that have one thing in common: the output is a DC voltage that is proportional to the applied pressure. Very few digital transducers exist. In fact, one that purportedly was "digital" was actually an analog transducer with a built-in analog-to-digital converter.

So we find that most of the instrumentation input data is analog in nature. What about output data displays? In many cases, perhaps most, we can use digital display devices for the output data. But sometimes we may require either an oscilloscope or strip-chart recorder for the display, and both of those are essentially analog displays. In still other cases, the output data produced by the computer will be in the form of an analog voltage or current that is used to drive some prime mover, such as a motor.

We have, therefore, established clearly a need for making both analog-to-digital (A/D) conversions and digital-to-analog conversions (DAC). Although the circuitry needed to accomplish these chores is beyond the scope of this chapter, we will discuss how these devices are interfaced with microcomputer and microprocessor instruments. Before using either A/D or DAC circuits, however, I recommend that you read a book on the subject; there are pitfalls to their use that are not immediately apparent.

Definitions

A *digital-to-analog converter* is a circuit or device that will produce either an output voltage or an output current that is proportional to the product of some reference voltage (or current) and a binary word applied to its inputs.

An *analog-to-digital converter* (A/D or ADC) is a circuit or device that will produce a binary digital word that is proportional to a reference voltage (or current) and an applied (unknown) voltage or current.

In addition to the definition for the DAC above, you will also see the term "multiplying DAC." This merely indicates that the DAC requires an external reference voltage (or current) source. The output, recall, is proportional to the *product* of the reference and the applied binary word. A nonmultiplying DAC will have the reference source built in.

20.4 DAC INTERFACING

Perhaps the easiest DAC interfacing chore is the connection of a DAC to a microcomputer output port of exactly the correct number of bits or less (there are problems when the DAC word length exceeds the microcomputer output port word length). This connection is shown in Figure 20-1. Many microcomputers have output ports in excess of their requirements, and it is often appropriate to use one of the "extras" for data conversion. In this case, it is merely necessary to connect the lines of the output port to their corresponding lines on the DAC. When a binary word is written to the output port, the word will be automatically connected to the input of the DAC, hence the DAC will produce an output voltage/current that is proportional to the binary word. This method, one of two basic methods, is called *I/O-based interfacing* of the DAC.

The method of Figure 20-1 assumes that there is an output port on the computer, and, further, that the output lines are *latched*. This is the

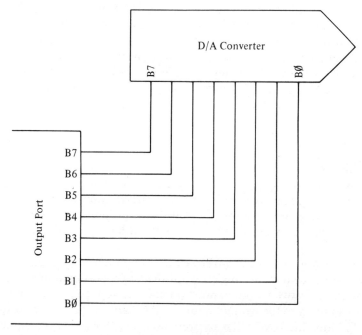

FIGURE 20-1. Interfacing DAC to output port.

Section 20.4 / DAC INTERFACING

situation that exists on most microcomputers. But oddly enough there are a few models on the market that use non-latched output ports, and these must somehow be accommodated by the instrumentation designer who (for some reason or another) must use them. Figure 20-2 shows a method for accommodating the non-latched microcomputer output port.

In this application we are limited to a single eight-bit output port, yet certain "housekeeping" chores must be performed, i.e., data strobes and data transfer. The circuit in Figure 20-2 uses a pair of 74100 dual quad latches. These ICs contain two independent, four-bit data latches. These

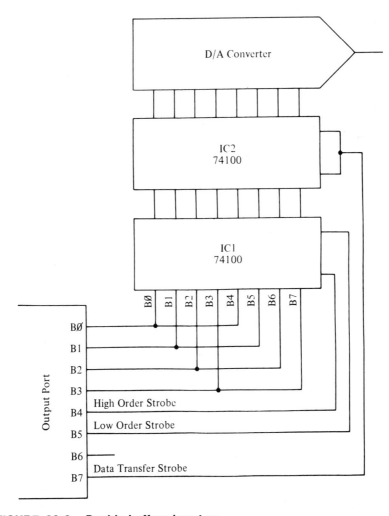

FIGURE 20-2. Double buffered system.

will transfer four-bit words on the input lines to the output lines whenever the strobe input is brought HIGH. By connecting the two strobe lines on a 74100 to a common line, we will make a single eight-bit data latch.

IC1 in Figure 20-2 is used with the two four-bit data latches independent. Notice that we connect both halves of IC1 to the same output port lines, i.e., B0 and B4 of the 74100 are connected to B0 of the output port, B1/B5 to bit B1 of the port, etc. We will need to write a program that will output half of the data word (i.e., four bits) at a time. For example, we could write the low order half byte to B0–B3 of the output port, and, in addition, make B5 HIGH. With B5 HIGH, the data on B0–B3 is input to B0–B3 of the 74100. Following that operation, B5 is made LOW again, and then the high order half byte of the data word is output to B0–B3 of the output port, and B4 of the output port is made HIGH. This condition turns on the high order strobe, so it will allow bits B0–B3 of the output port to be input to bits B4–B7 of the 74100. All through this sequence of operations, the DAC has been responding to the data on the outputs of the other 74100 device (IC2), so it will not undergo any changes. However, once we have loaded IC1 with the entire data word to be applied to the second 74100, then we can set B7 of the output port HIGH, causing the data that is latched onto the outputs of IC1 to be transferred to the outputs of IC2, hence to the DAC inputs. This method is called a *double-buffered latch*.

The interfacing methods thus far discussed in this chapter have been examples of I/O-based interfacing. In Figure 20-3 we see a method that could be either I/O-based, or, *memory-mapped* (also sometimes called *memory mapped-I/O*). Which it represents depends upon whether the OUT signal that controls the operation of the circuit is generated by an output command, or, a memory write command. The output device is the same 74100 that was used in previous examples. In this case, however, the input lines of the 74100 are connected directly to the data bus, which is eight bits long in most of the popular microcomputers. The output lines of the 74100 data latch are connected directly to the digital input lines of the digital-to-analog converter.

If this circuit is connected as an output port, then the OUT signal will be generated in response to an output command. Some microprocessors, such as the Z80 device, have separate I/O and memory write instructions. Others, such as the 68xx and 65xx series, use memory locations for the output port functions; these are called memory-mapped machines. In either case, the instruction being executed will place the output data on the data bus (D0–D7) and at the same time generate an appropriate device select (OUT) signal. When the OUT signal goes HIGH, then the 74100 will transfer data from the data bus to the DAC.

Not all DACs will require an external data latch to make these circuits work. An increasing number of devices are on the market that contain an internal data latch. One device, made by Ferranti Electric of England, has an internal latch that will also function as an internal eight-bit binary counter, depending upon the logic level applied to a control pin on the IC case.

Interfacing 10-, 12-, and 16-Bit DACs

Most microcomputers, and the microprocessor chips that form the computer, are eight-bit devices. An eight-bit data word will allow us to generate up to 256 discrete steps at the output of the DAC. In some cases, however, this does not result in sufficient resolution, so we must go to a DAC with a longer word length. For the sake of example, let us see exactly what the effect on resolution is when the data word length is manipulated.

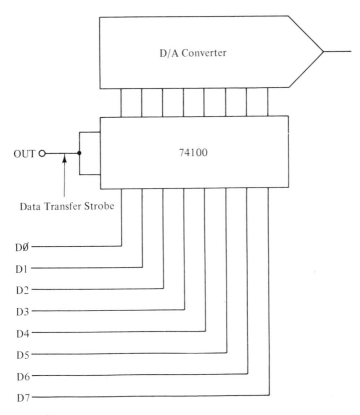

FIGURE 20-3. Interfacing DAC to data bus.

In the case of a system with a 10.00 volt fullscale output potential, the resolution will be 10/256 or about 40 mV per LSB change in the data word. In the case of a 10-bit DAC, however, the resolution is 10.00 V/(2^{10}), which is 10.00/1024, or about 9.8 mV. Similarly, 12-bit DACs produce a resolution of 2.4 mV and 16-bit DACs will resolve to 0.15 mV.

Obviously, extra bits on the DAC will result in more resolution (provided that the rest of the circuit is also upgraded—a 10-bit DAC connected to a reference source capable of supporting only an 8-bit DAC is still only an 8-bit DAC despite all the bits). But how do we make use of that extra precision in an eight-bit microcomputer. From a software point of view, we can use the double precision methods. Many of the popular microprocessor chip sets include double precision instructions. The Z80 device, for example, will allow certain instructions that make use of paired eight-bit registers to simulate the action of a 16-bit machine. In other microprocessor systems, we can write the software as double precision, treating successive eight-bit locations as if they were a single 16-bit location. Of course, this method will double our time requirements, but that is often a reasonable trade-off in instrument design.

The only question that remains, then, is how do we physically connect the 10- or more bit DAC to an eight-bit microcomputer. Figure 20-4 shows a general method that is applicable to 10- and 12-bit DACs, and by extension, to 16-bit DACs. In the 16-bit case, replace the four-bit registers (IC1 and IC3) with 74100 devices (Making IC1 through IC4 all 74100s). In this discussion we are going to use the Z80 terminology because it is among the easiest microprocessor chips to program for double precision operations. Keep in mind that the eight-bit microprocessor can only accommodate eight bits of data at a time because the data bus is only 8 bits in length. In a manner similar to Figure 20-2 we are going to *multiplex* the data (10-, 12- or 16-bits) onto the data bus in successive operations. To illustrate this process, let's assign three output port locations as follows:

175_{10} (AF$_{16}$)   Lower 8 bits of data to DAC
176_{10} (B0$_{16}$)   Higher (2, 4 or 8 bits) of data to the DAC
177_{10} (B1$_{16}$)   Control of output latch (IC3 and IC4)

This circuit (Figure 20-4) will first load the data into IC2, then load the remaining portion of the data into IC1, and then, simultaneously, transfer the data in IC1 and IC2 into IC3 and IC4, respectively. To do that job we need software that does the following:

1. Loads the CPU accumulator with the lower 8 bits of the 10-, 12- or 16-bit data word.

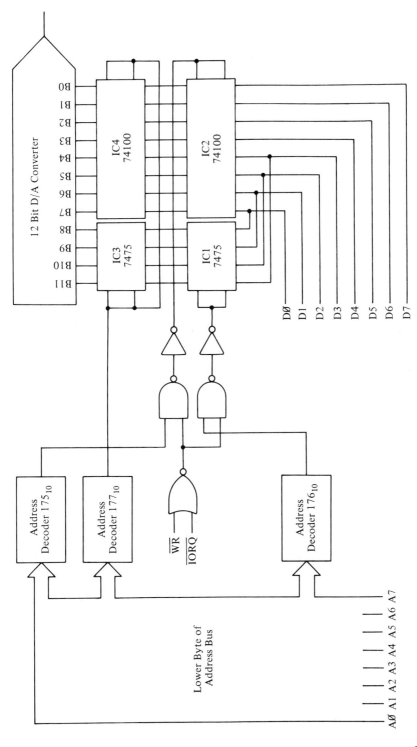

FIGURE 20-4. Interfacing eight-bit data bus to longer length DAC.

2. Outputs the lower 8 bits to port 175_{10} (i.e., loads IC2).
3. Loads the CPU accumulator with the higher 2, 4 or 8 bits of the data word.
4. Outputs the contents of the accumulator to port 176_{10} (i.e., loads IC1).
5. Generates a dummy output instruction to port 177_{10} in order to generate a device select pulse to that port. This operation will transfer data from IC1/IC2 to IC3/IC4—and thence to the DAC digital inputs.

It is important that this double-buffered approach be used because any other approach will result in erroneous outputs to the DAC. In some cases, a single-buffered method is used because (a) the DAC contains its own internal data latches or (b) because the DAC has a chip enable line that can be turned on and off as needed.

DAC Output Circuitry

Many of the lower cost DACs used in electronic instrumentation are current output devices. This means that the analog output of the DAC will be an electrical current, not a voltage. Unfortunately, most of the display devices and other electronic circuits that will receive the DAC output are voltage-oriented. We can overcome that problem by converting the DAC output current to a proportional voltage. Figure 20-5 shows a simple method involving an operational amplifier operated as an inverter. The inverting input is connected directly to the DAC current output and a feedback resistor scales the output voltage such that

$$E_o = I_o R$$

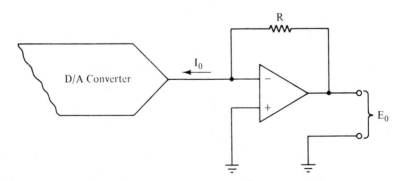

FIGURE 20-5. Creating voltage output for current output DACs.

where:

> E_o is the amplifier output potential, in volts
> I_o is the DAC output current in milliamperes
> R is the feedback resistor in kiloohms.

If the output voltage polarity is incorrect, then there are several steps that can be taken to correct the problem. One is to use a DAC with complementary current outputs, and then connect the output to the operational amplifier that produces the polarity required. Another tactic is to follow the amplifier in Figure 20-5 with an additional inverter stage, keeping in mind that the amplitude will be the product of the gains of both stages and the output current. Another tactic will be to use a noninverting follower instead of the inverting follower of Figure 20-5. Of course, it is still necessary to make the current-to-voltage conversion, but that can be done with a simple resistor. Connect the DAC output to the noninverting input of the operational amplifier. From this same point to ground connect a precision resistor. The voltage applied to the noninverting input of the operational amplifier will be the product of the current and the resistance.

One last problem involves filtering of the output signal. The DAC output signal will contain steps equal to the 1-LSB value of the DAC. If we were to connect a binary counter to the digital inputs of the DAC, then the result would be a staircase version of the old-fashioned sawtooth waveform. If we incorporate a low-pass filter between the DAC output and the display device, then this step-function is smoothed and will appear more like what we want to see. Select a filter cut-off frequency and slope that will smooth the signal without attenuating the main features of the waveform at the highest fundamental frequency of operation.

20.5 INTERFACING THE A/D CONVERTER

The A/D converter is the device that converts an analog input voltage or (sometimes) current into a proportional binary word that can be used in the computer. The subject of analog-to-digital conversion is beyond the scope of this chapter, but we will consider the A/D converter as a black box. Consider Figure 20-6. The typical A/D converter may have an analog input, a reference input (some use internal references), 8-, 10-, 12- or 16-bit binary digital output, and at least two service signals, *start* and *end of conversion* (EOC). Alternatively, some converters have a *status* line in lieu of the EOC line. The difference is that the EOC signal is a pulse that usually consumes but one cycle of the system clock (which may be either internal or external to the A/D), while the *status* signal will maintain one

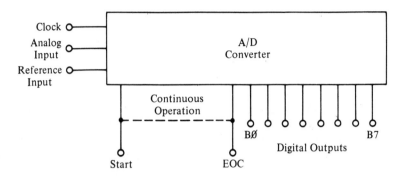

FIGURE 20-6. Basic A/D converter.

condition during the conversion process and a second condition when the data is valid. For example, one A/D converter has a *status* line that is HIGH when the conversion is taking place (which tells the computer and the outside world to not trust the output data) and LOW when the data is valid (indicating that the conversion is completed). The conversion time is the period from the issuance of a start pulse to the generation of the EOC pulse (or the status line equivalent). This is the fundamental limitation to the maximum speed of operation, hence the number of samples per second. Obviously, if it requires 100 mS to make a conversion, then it is unlikely that more than 10 samples per second are possible. The instrumentation designer must take into account the speed of the converter and the maximum sample rate required by the data being converted. If, for example, significant frequency components to 1000 Hz exist, then we must have (by Nyquist) at least 200 samples per second. This means a conversion not less often than every 0.5 milliseconds, or 500 microseconds. In that case, a low cost 40 mS A/D converter simply would not do the job.

The A/D converter will remain dormant, doing nothing but marking time, until a start pulse is received on the *start* input line. It will then make the conversion and then issue an EOC signal. The A/D converter then goes into a dormant state once more to await the arrival of a start pulse. In some cases, this is the most desirable mode of operation. In other cases, however, we will want to make continuous conversions. The quickest way to do that is to connect the EOC line to the start line, and let the EOC signal be the start signal for the next cycle. In some cases, the EOC is coupled to the start line via an OR gate, so that power-on pulses or manual start pulses may also be accommodated.

As in most microprocessor interfacing chores, there are two basic approaches that can be taken: I/O-based and memory-mapped. In the I/O-based system we use the input and output instruction of the micro-

Section 20.5 / INTERFACING THE A/D CONVERTER

processor (if any), and that requires some form of existing I/O port. If we are using a pre-assembled computer for our design, then it may be possible to use existing I/O ports.

We may also have to interface directly to the bus lines inside of the computer or directly to the microprocessor IC. In any event, we will have to provide some means to recognize the microprocessor's control signals and decode the port address.

In memory-mapped systems, on the other hand, we will treat the output of the A/D converter as another memory location in the total memory bank. It is the usual practice to assign the A/D converter a location in the upper 32K of the 64K total memory that is common on 8-bit machines. That convention is probably the result of few computers, either personal or "instrumentation" applications, that have a full complement of memory. In fact, there will be a significant number of instrumentation applications that can be filled with 1K, 2K, 4K, 8K, and 16K memory sizes. There are partisans to both I/O and memory-mapped approaches, but the logic seems to be on the side of the memory-mapped proponents in cases where the computer has less than the full 64K of memory.

In Figure 20-7 we see an I/O-based scheme that works, even though it is wasteful of I/O ports. This method might be preferred, however, in a case where you are using a device such as the 6502 microprocessor with a 6522 versatile interface adapter (VIA) and only need a single input port. This system requires one complete input port on the computer to accommodate the A/D output data, plus one bit each of another input and an

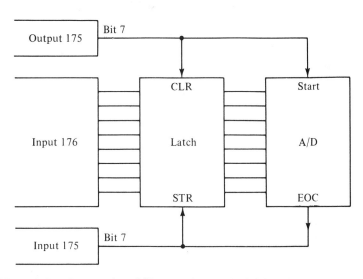

FIGURE 20-7. Connecting A/D converter to output port.

output port to accommodate the service signals from the A/D converter (*start* and *EOC*). In the example shown, we have (again) used port numbers 175 and 176. In this case, input port 176_{10} is used for the A/D output data, while bit no. 7 of input port 176_{10} is used to look for the EOC signal, and bit no. 7 of output port no. 175_{10} is used to generate the start signal. The A/D converter will initiate a conversion cycle whenever a HIGH appears on bit no. 7 of 175_{10}. The conversion will continue until complete, at which time a HIGH will be generated on the EOC line. This signal is applied to bit no. 7 of input port no. 175 (which tells the software that the data on 176_{10} input is good) and to the latch. In some cases, the latch will be activated by the EOC signal and then reset to LOW by the start signal.

A method that will accomplish the same job with less waste of I/O ports is shown in Figure 20-8. In this example we are multiplexing the port

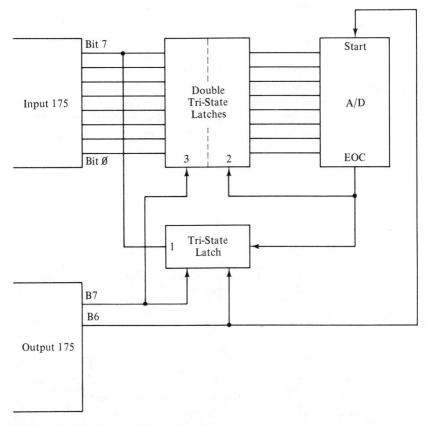

FIGURE 20-8. More efficient system.

Section 20.5 / INTERFACING THE A/D CONVERTER

so that it can be used for both service and data input functions. In this case, we use a single I/O port, no. 175_{10}. In order to initiate a conversion, we will output on port 175_{10} any binary word that has bit-6 HIGH and bit-7 LOW; e.g., 01000000_2 (40_{16}). If the A/D converter is edge-triggered, or if latch no. 1 is actually a Type-D flip-flop, then we can leave B6 HIGH. Otherwise, we will follow this operation with another write to 175_{10} with 00000000. In this situation we have the A/D converter making the conversion, and latch no. 1 looking for the EOC pulse. Note that the output of latch no. 1 is connected to one bit of input port 175_{10}. The computer software will loop looking for Bit-7 HIGH which indicates the A/D conversion is completed. The EOC pulse also cranks the output data into latch no. 2. After the program senses that the conversion is finished, it will output a HIGH on bit-7 of output port 175_{10}. This signal has the effect of turning on latch no. 3 (inputting the data) and turning off latch no. 1. In some cases, we might want to use separate bits for (first) turning off no. 1 and then turning on no. 3.

Figure 20-9 shows a simple method for connecting the A/D converter directly to the data bus. In this case, we are using the Z80 control signals for the example. For the example an I/O-based format is assumed. We could, however, make this a memory-mapped system through the simple expedient of using the memory request ($\overline{\text{MREQ}}$) signal instead of the Input Output Request ($\overline{\text{IORQ}}$) signal, as shown. The A/D converter selected (or designed) will have latched output lines (to keep the data valid as long as needed for the computer to respond) and a status line to indicate when the data is valid.

In the Z80 system, the OUT operation requires two signals: $\overline{\text{IORQ}}$ and the *write* ($\overline{\text{WR}}$); these indicate an *I/O write* which is an output. Similarly, an input operation requires $\overline{\text{IORQ}}$ and $\overline{\text{RD}}$; these form the *I/O read* signal. Gate no. 1 forms the start circuitry for the A/D converter. Since both of the control signals ($\overline{\text{IORQ}}$ and $\overline{\text{WR}}$) are active-LOW, we will use a NOR gate as an NAND-negative-logic gate. In this case, both inputs have to be LOW for the output to be HIGH. When the computer program executes a proper output command, then the output of gate no. 1 will go HIGH, initiating a start of conversion for the A/D. Immediately thereafter, the computer will loop, looking for a valid input. Gate no. 2 creates the I/O read, while gate no. 3 prevents it from occurring until the data is valid. During the conversion period the status line will be LOW, so gate no. 3 will not produce the required LOW output.

None of the A/D converter scenarios presented thus far are particularly appealing; all suffer from the necessity of tying up the computer looking for some input condition. In Figure 20-10 we see a system that takes advantage of the interrupt capability of the machine. A computer will respond to the interrupt by temporarily putting aside the program

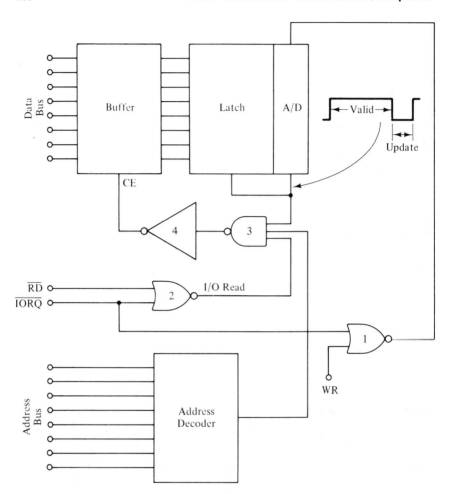

FIGURE 20-9. Interfacing A/D to data bus.

being executed and switching to a subroutine that will service the interrupt. Let's take an instrumentation example of how this might be used. Assume that we are going to build a data logger and averager for, say, an electroencephalograph (EEG) evoked potentials computer. The A/D converter will be continuously making conversions of the patient's brain wave signal. In the case of Figure 20-10, the converter is connected in the continuous, or asynchronous mode. We will input a data point, store it in memory, and then average it with the number of data points already received. Typically, we could add the "to be averaged" number in a 16-bit register pair (or two sequential memory locations), and keep a software

Section 20.5 / INTERFACING THE A/D CONVERTER

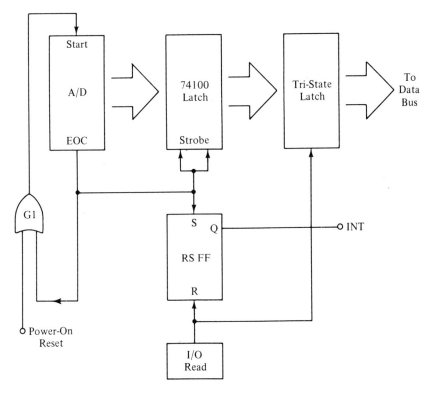

FIGURE 20-10. A/D driving interrupt line.

(or hardware in some computers) count of the total number of samples. For a given time, the value in the averager will accumulate. When the "run" is completed, then the total accumulated value will be divided by the value of the counter. Of course, for a real evoked potentials computer, we would need a separate "averager" location for each data point.[1] The purpose of using the interrupt in this system is to allow the computer time to do the averaging and other chores while the data converter is making the conversion. When a new conversion is ready, the A/D circuit will interrupt the computer, input the new data, and then go on to make a newer yet conversion. During the time between the input operations the computer can do its "number crunching thing."

The start signal is coupled through a two-input OR gate. This arrangement allows us to couple either the *EOC* signal or the *power-on*

[1] See *Introduction to Biomedical Equipment Technology* by J.J. Carr and J.M. Brown, New York: John Wiley & Sons, 1981.

reset signal to the start terminal of the A/D converter. When the conversion is completed the EOC pulse is generated. This pulse will (a) input the A/D data to the 74100 latch, and (b) set the RS flip-flop (making Q = HIGH). If the computer wants to see an INT signal, then use the Q output of the RS FF. Otherwise, if an $\overline{\text{INT}}$ is needed (most often the case), then use the $\overline{\text{Q}}$ output of the RS FF.

When the computer finishes the execution of the instruction during which the INT or $\overline{\text{INT}}$ signal is received, it will store the location of the program counter on a stack some place in memory, and then switch to executing the interrupt subroutine. In the case of the A/D converter interrupt, that program will generate an I/O read operation. This means that a device select pulse for the I/O port servicing the A/D will be generated, so it may be used to (a) reset the RS FF, and (b) turn on the tri-state latch that places the A/D data (now on the output of the 74100 latch) onto the data bus of the computer.

As an alternative, we can connect the start line of the A/D to a second I/O read signal so that the conversion will start only under software control.

20.6 SUMMARY

1. Both A/D and DACs can be connected to the microcomputer using either memory-mapped or I/O-based methods.
2. The control of the A/D can be either asynchronous or under total software control.

20.7 QUESTIONS

20-1. Define (a) memory-mapped and (b) I/O-based systems.
20-2. What must sometimes be done at the output of a DAC to make the signal smooth?
20-3. How may the output voltage of a DAC be scaled?
20-4. How may a current-output DAC be used with voltage-input displays?
20-5. What is the fundamental limitation that determines the highest sample rate attainable on an A/D converter?

21
Software Data Conversion

21.1 OBJECTIVES

1. To write software binary ramp data conversion programs.
2. To write software successive approximation data conversion programs.
3. Write a program for a tracking A/D converter.
4. Build a digital *Sample & Hold*.

21.2 SELF-EVALUATION QUESTIONS

Before studying the material in this chapter, try answering the questions given below. These questions test your knowledge of the subject matter. If you cannot answer any particular question, then place a check mark beside it and look for the answer as you read the text.

1. Which A/D conversion methods are normally used in software A/D converters?
2. Why is a digital Sample & Hold often superior to an analog Sample & Hold circuit?
3. How many times must a binary ramp routine execute to make a fullscale 8-bit conversion?
4. How much time must a successive approximation routine have to make a fullscale 8-bit conversion?

21.3 INTRODUCTION

A great advantage of the digital computer is the ability to substitute *software* routines (i.e., programming) for *hardware* logic devices. This ability becomes especially appealing if the software logic can be performed during otherwise "lost" time on the computer. The software approach is appealing because it is both more reliable and more versatile than hardware. You can, for example, change the program almost at will by replacing a read only memory (ROM) chip. To make any major change in hardware requires a lot more effort, including the possibility of redesigning the entire printed circuit board or chassis wiring schemes. Even relatively minor hardware changes can result in massive changes in the printed circuit board.

You gain flexibility also because the computer is a "universal analytical engine." You can make a single standard single-board computer or a universal system of boards that mate to a common mother board or bus and can be used in many different electronic instruments and control systems. Some manufacturers offer single-board computers based on popular microprocessor chips. These boards have limited RAM, and usually about 2 Kbytes of ROM on-board. By using the data, address, and I/O lines, a designer can press the same board into service in a dozen different applications. The same computer printed circuit board, loaded with different programs in ROM, will perform these different functions.

Two A/D conversion techniques are most suitable for software implementation: binary ramp and successive approximation. This chapter will discuss both methods. You will receive some insight as to the type of programming needed and the type of external circuitry required to support the two methods. Interestingly enough, almost identical external circuitry will suffice for both binary ramp and SA conversion routines.

21.4 SOFTWARE RAMP CONVERTERS

The binary ramp type of A/D converter can be built using the software approach with very little external hardware. Figure 21-1 shows the connections required to use microcomputer I/O ports for this purpose. The digital inputs of the DAC are connected to the latched output port of the microcomputer. If there are no available I/O ports, then make one using one of the methods given in Chapter 10. The additional integrated circuits can be built right on the same board as the A/D converter components.

The output of the comparator is connected to one bit of a microcomputer input port. One input of the comparator is connected to the output of the DAC, while the other comparator input is connected to the analog input voltage being measured.

The program that operates this A/D converter must first initialize

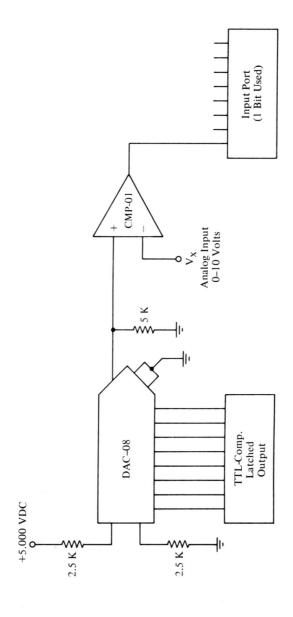

FIGURE 21-1. Connections for software A/D converter.

some internal counter to either zero or full scale and then create a binary ramp output. This job is accomplished by outputting the counter to the contents of the DAC and then incrementing the counter. After each output to the DAC, the input bit used as the comparator flag is tested. If the bit is found to indicate that the input voltage is not yet equal to the DAC output voltage, then the computer will jump back and increment the counter and output the new value to the DAC. This loop continues until the comparator output drops LOW, indicating that the analog input voltage is equal to the DAC output voltage. The state of the counter that caused this condition is then used to represent the analog input voltage. To briefly summarize the process:

Assume that we have a circuit in which a HIGH output of the comparator means that the analog input voltage is not equal to the DAC output voltage. The computer will loop through the process given below until it finds that the comparator output is LOW (indicating equality of the analog input voltage and the DAC output voltage).

1. The counter is initialized to either OO or FF (Hex).
2. The test bit (output of the comparator) is tested for LOW.
3. If the test bit turned out to be HIGH, then the program will increment (or decrement if FF was used initially) the counter, and the result output to the DAC.
4. If the test bit was LOW, then the program moves the contents of the counter to the location in memory designed to hold the analog input value.

In some cases, there might also be some other functions that the program would have to perform. For example, it might be desirable to issue a pulse, or a data word output, that tells the outside world (or another instrument) that the A/D conversion process has started. This job could be accomplished by outputting a HIGH to one bit of some specific output port. In some cases, we will want an end-of-conversion (EOC) pulse rather than a status signal. This job is best performed by making the computer think it is a monostable multivibrator. In almost all computers, we can do this by outputting a HIGH to one bit of an output port, then looping through a do-nothing program for a specified number of microseconds, and then setting the same bit LOW. Some computers have a timer that will operate in the monostable mode (see the 6522 in the 650x family). Writing a monostable multivibrator program is a good exercise for new microcomputerists.

21.5 SOFTWARE SUCCESSIVE APPROXIMATION A/D CONVERSION

The successive approximation method was discussed in an earlier chapter. This simple technique sets the MSB of an output register HIGH, thereby setting the output of the DAC connected to that port to midscale. The unknown voltage is tested against this level, and the subsequent action taken depends upon the results. If the trial output value is lower than the unknown, then the bit is set permanently HIGH, and the next trial is made. This trial will involve the next most significant bit (MSB-1). If, on the other hand, the trial output value is too high, then the MSB is reset to LOW, and the MSB-1 bit is tried. This procedure is continued until all of the bits have been tried. This logic is implemented in hardware with logic devices, and at least two manufacturers offer IC successive approximation registers in 8, 10, and 12-bit lengths. Since *logic* is involved, however, we can also implement the process in software.

The very same external circuit used in the binary ramp method (section 21.4) will also work with the successive approximation methods. We may also use a *memory-mapped* method such as Figure 21-2 for these jobs. The circuit of Figure 21-2 shows a *Precision Monolithics, Inc.* DAC-08 digital-to-analog converter interfaced with the Intel 8080A microprocessor chip. This particular system is memory-mapped into the upper 32K of the available 64K memory space. Bit $A15$, the most significant bit of the 16-bit address bus, is used to turn on the circuit. The digital inputs of the DAC-08 are connected to bits A0 through A7 of the address bus. The output of the DAC-08 is connected to one input of a voltage comparator. This circuit uses the comparator operated in the current mode, which allows us to take better advantage of the DAC-08's inherent speed. Notice that the I_o output of the DAC-08 is connected to the analog input of the circuit. This is done so that the analog input impedance remains constant for different values of input voltage.

This circuit uses the DAC-08 as if it were a 256 × 1 bit memory location. Note that the selected DAC and the comparator *both* have a settling time specification that are sufficient to follow the 8080A microcomputer, so no "NOP" delay-loop instructions or wait states are needed in the program. The following is a sample program in 8080A assembler lanquage, and will operate the circuit in Figure 21-2 in a successive approximation routine. Figure 21-3 is a simple flow diagram of this program.

```
START:   LXI   B, 08000H   ; LOAD MSP IN B, CLEAR C
         MOV A, B          ; MSE TO ACC
         MOV H, A          ; SET MEM/MAP I/O
```

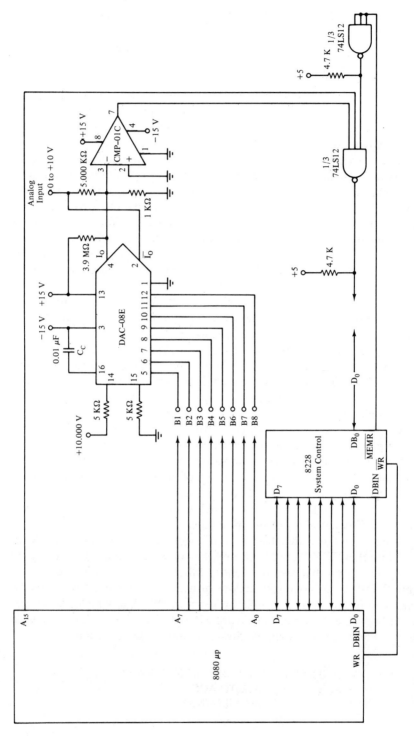

FIGURE 21-2. DAC-08 software A/D converter.

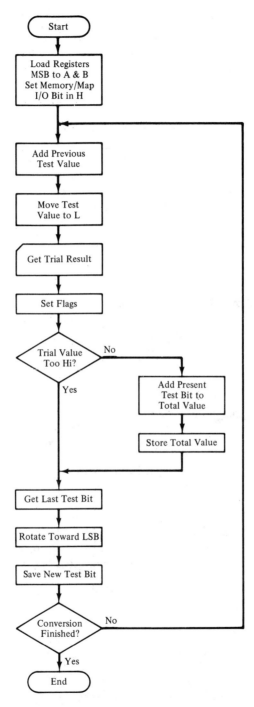

FIGURE 21-3. Software flowchart for A/D conversion routine.

```
TEST:   ORA C           ; ADD LAST TEST VALUE
        MOV L, A        ; MOVE PRESENT TEST TO L
        MOV A, M        ; GET COMP OUTPUT
        ANA A           ; SET FLAGS
        JPO TOOHI       ; DISCARD PRESENT TEST BIT
        MOV A, B        ; GET PRESENT TEST BIT
        ORA C           ; ADD TOTAL SO FAR
        MOV C, A        ; SAVE TOTAL
TOOHI:  MOV A, B        ; GET LAST TEST BIT
        RAR             ; ROTATE TOWARD LSB
        MOV B, A        ; SAVE NEW TEST FIT
        JNC TEST        ; JUMP IF NOT FINISH
        END             ; FINAL VALUE IS IN C
```

SOURCE: Precision Monolithics, Inc.

21.6 SOFTWARE DATA CONVERSION EXERCISES

These exercises are designed to familiarize you with some practical applications. They will run on most microcomputers, but were specifically selected for users of 6502/6522 based machines such as the AIM-65, KIM-1, and SYM-1.

21.1. Use the circuit of Figure 21-1 to build a binary ramp A/D converter. Write a program that will: (a) initiate the conversion process, (b) perform the conversion, (c) issue a status level (HIGH for *busy* and LOW for *data ready*) to the outside world, and (d) store the converted value at location 000A (Hex). *Note:* Some designers will use one bit of one of the two input ports to test the comparator output, while others will use the counter/timer section of the VIA chip (6522).

21.2. Write a program that will use the program of Exercise 21.1 to store 256 successive samples in memory.

21.3. Write programs for Exercises 21.1 and 21.3, using 360 different samples for Exercise 21.2 instead of 256. Then write a program that will store a 10 Hz sinewave in memory. Include methods for determining when the individual sinewave starts and stops.

21.4. Write the programs of Exercises 21.1 through 21.3, and then, after the sinewave is stored in memory, be able to output it through the same DAC for display on an oscilloscope screen or on a strip chart recorder if your lab is so equipped. *Note:* Some designers prefer to initialize the DAC at 00H and then ramp upwards. Others start at FFH and ramp downwards. Discuss these two approaches and give reasons why one might be superior to the other in some cases.

Section 21.6 / SOFTWARE DATA CONVERSION EXERCISES

21.5. Use the successive approximation algorithm to perform any of the above exercises, as assigned.

21.6. A *tracking* A/D converter makes continuous conversions and will input new data whenever there is a change. Write a program based on either type of software A/D converter that will track changes in the analog input voltage value. Store the value in a memory location in page zero or in one of the index registers. *Problem:* Can you see any possibility for hysteresis in the data? In other words, will the new value be different if it is approached from below than if from above? Explain this phenomenon, how it works, and propose a solution that will always yield the same answer, regardless of the direction from which the new value is approached. Incorporate your solution into the program.

21.7. Build a digital *Sample & Hold* based on (a) the binary ramp method or (b) the successive approximation method. Explain why the digital Sample & Hold circuit might be superior to the analog (operational amplifier, switch, and capacitor) version normally used in some instruments.

21.8. How would you modify the circuit of Figure 21-1 to prevent errors caused by changes in the analog input voltage during the time when the conversion is taking place. Is this problem greater in either of the two conversion methods. If so, then why?

Appendix A
Z80 Instructions Sorted By Op-Code

OBJ CODE	SOURCE STATEMENT	OBJ CODE	SOURCE STATEMENT	OBJ CODE	SOURCE STATEMENT
00	NOP	282E	JR Z, DIS	50	LD D, B
018405	LD BC, NN	29	ADD HL, HL	51	LD D, C
02	LD (BC), A	2A8405	LD (HL), (NN)	52	LD D, D
03	INC BC	2B	DEC HL	53	LD D, E
04	INC B	2C	INC L	54	LD D, H
05	DEC B	2D	DEC L	55	LD D, L
0620	LD B, N	2E20	LD L, N	56	LD D, (HL)
07	RLCA	2F	CPL	57	LD D, A
08	EX AF, AF'	302E	JR NC, DIS	58	LD E, B
09	ADD HL, BC	318405	LD SP, NN	59	LD E, C
0A	LD A, (BC)	328405	LD (NN), A	5A	LD E, D
0B	DEC BC	33	INC SP	5B	LD E, E
0C	INC C	34	INC (HL)	5C	LD E, H
0D	DEC C	35	DEC (HL)	5D	LD E, L
0E20	LD C, N	3620	LD (HL), N	5E	LD E, (HL)
0F	RRCA	37	SCF	5F	LD E, A
102E	DJNZ DIS	382E	JR C, DIS	60	LD H, B
118405	LD DE, NN	39	ADD HL, SP	61	LD H, C
12	LD (DE), A	3A8405	LD A, (NN)	62	LD H, D
13	INC DE	3B	DEC SP	63	LD H, E
14	INC D	3C	INC A	64	LD H, H
15	DEC D	3D	DEC A	65	LD H, L
1620	LD D, N	3E20	LD A, N	66	LD H, (HL)
17	RLA	3F	CCF	67	LD H, A
182E	JR DIS	40	LD B, B	68	LD L, B
19	ADD HL, DE	41	LD B, C	69	LD L, C
1A	LD A, (DE)	42	LD B, D	6A	LD L, D
1B	DEC DE	43	LD B, E	6B	LD L, E
1C	INC E	44	LD B, H, NN	6C	LD L, H
1D	DEC E	45	LD B, L	6D	LD L, L
1E20	LD E, N	46	LD B, (HL)	6E	LD L, (HL)
1F	RRA	47	LD B, A	6F	LD L, A
202E	JR NZ, DIS	48	LD C, B	70	LD (HL), B
218405	LD HL, NN	49	LD C, C	71	LD (HL), C
228405	LD (NN), HL	4A	LD C, D	72	LD (HL), D
23	INC HL	4B	LD C, E	73	LD (HL), E
24	INC H	4C	LD C, H	74	LD (HL), H
25	DEC H	4D	LD C, L	75	LD (HL), L
2620	LD H, N	4E	LD C, (HL)	76	HALT
27	DAA	4F	LD C, A	77	LD (HL), A

Z80 INSTRUCTIONS SORTED BY OP-CODE / Appendix A

OBJ CODE	SOURCE STATEMENT	OBJ CODE	SOURCE STATEMENT	OBJ CODE	SOURCE STATEMENT
78	LD A, B	A0	AND B	C8	RET Z
79	LD A, C	A1	AND C	C9	RET
7A	LD A, D	A2	AND D	CA8405	JP Z, NN
7B	LD A, E	A3	AND E	CC8405	CALL Z, NN
7C	LD A, H	A4	AND H	CD8405	CALL NN
7D	LD A, L	A5	AND L	CE20	ADC A, N
7E	LD A, (HL)	A6	AND (HL)	CF	RST 8
7F	LD A, A	A7	AND A	D0	RET NC
80	ADD A, B	A8	XOR B	D1	POP DE
81	ADD A, C	A9	XOR C	D28405	JP NC, NN
82	ADD A, D	AA	XOR D	D320	OUT (N), A
83	ADD A, E	AB	XOR E	D48405	CALL NC, NN
84	ADD A, H	AC	XOR H	D5	PUSH DE
85	ADD A, L	AD	XOR L	D620	SUB N
86	ADD A, (HL)	AE	XOR (HL)	D7	RST 10H
87	ADD A, A	AF	XOR A	D8	RET C
88	ADC A, B	B0	OR B	D9	EXX
89	ADC A, C	B1	OR C	DA8405	JP C, NN
8A	ADC A, D	B2	OR D	DB20	IN A, (N)
8B	ADC A, E	B3	OR E	DC8405	CALL C, N
8C	ADC A, H	B4	OR H	DE20	SBC A, N
8D	ADC A, L	B5	OR L	DF	RST 18H
8E	ADC A, (HL)	B6	OR (HL)	E0	RET PO
8F	ADC A, A	B7	OR A	E1	POP HL
90	SUB B	B8	CP B	E28405	JP PO, NN
91	SUB C	B9	CP C	E3	EX (SP), HL
92	SUB D	BA	CP D	E48405	CALL PO, NN
93	SUB E	BB	CP E	E5	PUSH HL
94	SUB H	BC	CP H	E620	AND N
95	SUB L	BD	CP L	E7	RST 20 H
96	SUB (HL)	BE	CP (HL)	E8	RET PE
97	SUB A	BF	CP A	E9	JP (HL)
98	SBC A, B	C0	RET NZ	EA8405	JE PE NN
99	SBC A, C	C1	POP BC	EB	EX DE, HL
9A	SBC A, D	C28405	JP NZ, NN	EC8405	CALL PE, NN
9B	SBC A, E	C38405	JP NN	EE20	XOR N
9C	SBC A, H	C48405	CALL NZ, NN	EF	RST 28H
9D	SBC A, L	C5	PUSH BC	F0	RET P
9E	SBC A, (HL)	C620	ADD A, N	F1	POP AF
9F	SBC A, A	C7	RST 0	F28405	JP P, NN

Z80 INSTRUCTIONS SORTED BY OP-CODE / Appendix A

OBJ CODE	SOURCE STATEMENT	OBJ CODE	SOURCE STATEMENT	OBJ CODE	SOURCE STATEMENT
F3	DI	CB1C	RR H	CB4C	BIT 1, H
F48405	CALL P, NN	CB1D	RR L	CB4D	BIT 1, L
F5	PUSH AF	CB1E	RR (HL)	CB4E	BIT 1, (HL)
F620	OR N	CB1F	RR A	CB4F	BIT 1, A
F7	RST 30H	CB20	SLA B	CB50	BIT 2, B
F8	RET M	CB21	SLA C	CB51	BIT 2, C
F9	LD SP, HL	CB22	SLA D	CB52	BIT 2, D
FA8405	JP M, NN	CB23	SLA E	CB53	BIT 2, E
FB	EI	CB24	SLA H	CB54	BIT 2, H
FC8405	CALL M, NN	CB25	SLA L	CB55	BIT 2, L
FE20	CP N	CB26	SLA (HL)	CB56	BIT 2, (HL)
FF	RST 38H	CB27	SLA A	CB57	BIT 2, A
CB00	RLC B	CB28	SRA B	CB58	BIT 3, B
CB01	RLC C	CB29	SRA C	CB59	BIT 3, C
CB02	RLC D	CB2A	SRA D	CB5A	BIT 3, D
CB03	RLC E	CB2B	SRA E	CB5B	BIT 3, E
CB04	RLC H	CB2C	SRA H	CB5C	BIT 3, H
CB05	RLC L	CB2D	SRA L	CB5D	BIT 3, L
CB06	RLC (HL)	CB2E	SRA (HL)	CB5E	BIT 3, (HL)
CB07	RLC A	CB2F	SRA A	CB5F	BIT 3, A
CB08	RRC B	CB38	SRL B	CB60	BIT 4, B
CB09	RRC C	CB39	SRL C	CB61	BIT 4, C
CB0A	RRC D	CB3A	SRL D	CB62	BIT 4, D
CB0B	RRC E	CB3B	SRL E	CB63	BIT 4, E
CB0C	RRC H	CB3C	SRL H	CB64	BIT 4, H
CB0D	RRC L	CB3D	SRL L	CB65	BIT 4, L
CB0E	RRC (HL)	CB3E	SRL (HL)	CB66	BIT 4, (HL)
CB0F	RRC A	CB3F	SRL A	CB67	BIT 4, A
CB10	RL B	CB40	BIT 0, B	CB68	BIT 5, B
CB11	RL C	CB41	BIT 0, C	CB69	BIT 5, C
CB12	RL D	CB42	BIT 0, D	CB6A	BIT 5, D
CB13	RL E	CB43	BIT 0, E	CB6B	BIT 5, E
CB14	RL H	CB44	BIT 0, H	CB6C	BIT 5, H
CB15	RL L	CB45	BIT 0, L	CB6D	BIT 5, L
CB16	RL (HL)	CB46	BIT 0, (HL)	CB6E	BIT 5, (HL)
CB17	RL A	CB47	BIT 0, A	CB6F	BIT 5, A
CB18	RR B	CB48	BIT 1, B	CB70	BIT 6, B
CB19	RR C	CB49	BIT 1, C	CB71	BIT 6, C
CB1A	RR D	CB4A	BIT 1, D	CB72	BIT 6, D
CB1B	RR E	CB4B	BIT 1, E	CB73	BIT 6, E

Z80 INSTRUCTIONS SORTED BY OP-CODE / Appendix A

OBJ CODE	SOURCE STATEMENT	OBJ CODE	SOURCE STATEMENT	OBJ CODE	SOURCE STATEMENT
CB74	BIT 6, H	CB9C	RES 3, H	CBC4	SET 0, H
CB75	BIT 6, L	CB9D	RES 3, L	CBC5	SET 0, L
CB76	BIT 6, (HL)	CB0E	RES 3, (HL)	CBC6	SET 0, (HL)
CB77	BIT 6, A	CB9F	RES 3, A	CBC7	SET 0, A
CB78	BIT 7, B	CBA0	RES 4, B	CBC8	SET 1, B
CB79	BIT 7, C	CBA1	RES 4, C	CBC9	SET 1, C
CB7A	BIT 7, D	CBA2	RES 4, D	CBCA	SET 1, D
CB7B	BIT 7, E	CBA3	RES 4, E	CBCB	SET 1, E
CB7C	BIT 7, H	CBA4	RES 4, H	CBCC	SET 1, H
CB7D	BIT 7, L	CBA5	RES 4, L	CBCD	SET 1, L
CB7E	BIT 7, (HL)	CBA6	RES 4, (HL)	CBCE	SET 1, (HL)
CB7F	BIT 7, A	CBA7	RES 4, A	CBCF	SET 1, A
CB80	RES 0, B	CBA8	RES 5, B	CBD0	SET 2, B
CB81	RES 0, C	CBA9	RES 5, C	CBD1	SET 2, C
CB82	RES 0, D	CBAA	RES 5, D	CBD2	SET 2, D
CB83	RES 0, E	CBAB	RES 5, E	CBD3	SET 2, E
CB84	RES 0, H	CBAC	RES 5, H	CBD4	SET 2, H
CB85	RES 0, L	CBAD	RES 5, L	CBD5	SET 2, L
CB86	RES 0, (HL)	CBAE	RES 5, (HL)	CBD6	SET 2, (HL)
CB87	RES 0, A	CBAF	RES 5, A	CBD7	SET 2, A
CB88	RES 1, B	CBB0	RES 6, B	CBD8	SET 3, B
CB89	RES 1, C	CBB1	RES 6, C	CBD9	SET 3, C
CB8A	RES 1, D	CBB2	RES 6, D	CBDA	SET 3, D
CB8B	RES 1, E	CBB3	RES 6, E	CBDB	SET 3, E
CB8C	RES 1, H	CBB4	RES 6, H	CBDC	SET 3, H
CB8D	RES 1, L	CBB5	RES 6, L	CBDD	SET 3, L
CB8E	RES 1, (HL)	CBB6	RES 6, (HL)	CBDE	SET 3, (HL)
CB8F	RES 1, A	CBB7	RES 6, A	CBDF	SET 3, A
CB90	RES 2, B	CBB8	RES 7, B	CBE0	SET 4, B
CB91	RES 2, C	CBB9	RES 7, C	CBE1	SET 4, C
CB92	RES 2, D	CBBA	RES 7, D	CBE2	SET 4, D
CB93	RES 2, E	CBBB	RES 7, E	CBE3	SET 4, E
CB94	RES 2, H	CBBC	RES 7, H	CBE4	SET 4, H
CB95	RES 2, L	CBBD	RES 7, L	CBE5	SET 4, L
CB96	RES 2, (HL)	CBBE	RES 7, (HL)	CBE6	SET 4, (HL)
CB97	RES 2, A	CBBF	RES 7, A	CBE7	SET 4, A
CB98	RES 3, B	CBC0	SET 0, B	CBE8	SET 5, B
CB99	RES 3, C	CBC1	SET 0, C	CBE9	SET 5, C
CB9A	RES 3, D	CBC2	SET 0, D	CBEA	SET 5, D
CB9B	RES 3, E	CBC3	SET 0, E	CBEB	SET 5, E

OBJ CODE	SOURCE STATEMENT	OBJ CODE	SOURCE STATEMENT	OBJ CODE	SOURCE STATEMENT
CBEC	SET 5, H	CBF1	SET 6, C	CBF6	SET 6, (HL)
CBED	SET 5, L	CBF2	SET 6, D	CBF7	SET 6, A
CBEE	SET 5, (HL)	CBF3	SET 6, E	CBF8	SET 7, B
CBEF	SET 5, A	CBF4	SET 6, H	CBF9	SET 7, C
CBF0	SET 6, B	CBF5	SET 6, L	CBFA	SET 7, D

Appendix B
Z80 Instructions Sorted By Mnemonic

Z80 INSTRUCTIONS SORTED BY MNEMONIC / Appendix B

OBJ CODE	SOURCE STATEMENT	OBJ CODE	SOURCE STATEMENT	OBJ CODE	SOURCE STATEMENT
8E	ADC A, (HL)	DD29	ADD IX, IX	CB49	BIT 1, C
DD8E05	ADC A, (IX + d)	DD39	ADD IX, SP	CB4A	BIT 1, D
FD8E05	ADC A, (IY + d)	FD09	ADD IY, BC	CB4B	BIT 1, E
8F	ADC A, A	FD19	ADD IY, DE	CB4C	BIT 1, H
88	ADC A, U	FD29	ADD IY, IY	CB4D	BIT 1, L
89	ADC A, C	FD39	ADD IY, SP	CB56	BIT 2, (HL)
8A	ADC A, D	A6	AND (HL)	DDCB0556	BIT 2, (IX + d)
8B	ADC A, E	DDA605	AND (IX + d)	FDCB0556	BIT 2, (IY + d)
8C	ADC A, H	FDA605	AND (IY + d)	CB57	BIT 2, A
8D	ADC A, L	A7	AND A	CB50	BIT 2, B
CE20	ADC A, N	A0	AND B	CB51	BIT 2, C
ED4A	ADC HL, BC	A1	AND C	CB52	BIT 2, D
ED5A	ADC HL, DE	A2	AND D	CB53	BIT 2, E
ED6A	ADC HL, HL	A3	AND E	CB54	BIT 2, H
ED7A	ADC HL, SP	A4	AND H	CB55	BIT 2, L
86	ADD A, (HL)	A5	AND L	CB5E	BIT 3, (HL)
DD8605	ADD A, (IX + d)	E620	AND N	DDCB055E	BIT 3, (IX + d)
FD8605	ADD A, (IY + d)	CB46	BIT 0, (HL)	FDCB055E	BIT 3, (IY + d)
87	ADD A, A	DDCB0546	BIT 0, (IX + d)	CB5F	BIT 3, A
80	ADD A, B	FDCB0546	BIT 0, (IY + d)	CB58	BIT 3, B
81	ADD A, C	CB47	BIT 0, A	CB59	BIT 3, C
82	ADD A, D	CB40	BIT 0, B	CB5A	BIT 3, D
83	ADD A, E	CB41	BIT 0, C	CB5B	BIT 3, E
84	ADD A, H	CB42	BIT 0, D	CB5C	BIT 3, H
85	ADD A, L	CB43	BIT 0, E	CB5D	BIT 3, L
C620	ADD A, N	CB44	BIT 0, H	CB66	BIT 4, (HL)
09	ADD HL, BC	CB45	BIT 0, L	DDCB0566	BIT 4, (IX + d)
19	ADD HL, DE	CB4E	BIT 1, (HL)	FDCB0566	BIT 4, (IY + d)
29	ADD HL, HL	DDCB054E	BIT 1, (IX + d)	CB67	BIT 4, A
39	ADD HL, SP	FDCB054E	BIT 1, (IY + d)	CB60	BIT 4, B
DD09	ADD IX, BC	CB4F	BIT 1, A	CB61	BIT 4, C
DD19	ADD IX, DE	BC48	BIT 1, B	CB62	BIT 4, D

Z80 INSTRUCTIONS SORTED BY MNEMONIC / Appendix B

OBJ CODE	SOURCE STATEMENT	OBJ CODE	SOURCE STATEMENT	OBJ CODE	SOURCE STATEMENT
CB63	BIT 4, E	BF	CP A	ED48	IN C, (C)
CB64	BIT 4, H	B8	CP B	ED50	IN D, (C)
CB65	BIT 4, L	B9	CP C	ED58	IN E, (C)
CB6E	BIT 5, (HL)	BA	CP D	ED60	IN H, (C)
DDCB056E	BIT 5, (IX + d)	BB	CP E	ED68	IN L, (C)
FDCB056E	BIT 5, (IY + d)	BC	CP H	34	INC (HL)
CB6F	BIT 5, A	BD	CP L	DD3405	INC (IX + d)
CB68	BIT 5, B	FE20	CP N	FD3405	INC (IY + d)
CB69	BIT 5, C	EDA9	CPD	3C	INC A
CB6A	BIT 5, D	ED89	CPDR	04	INC B
CB6B	BIT 5, E	EDA1	CPI	03	INC BC
CB6C	BIT 5, H	EDB1	CPIR	0C	INC C
CB6D	BIT 5, L	2F	CPL	14	INC D
CB76	BIT 6, (HL)	27	DAA	13	INC DE
DDCB0576	BIT 6, (IX + d)	35	DEC (HL)	1C	INC E
FDCB0578	BIT 5, (IY + d)	DD3505	DEC (IX + d)	24	INC H
CB77	BIT 6, A	FD3505	DEC (IY + d)	23	INC HL
CB70	BIT 6, B	3D	DEC A	DD23	INC IX
CB71	BIT 6, C	05	DEC B	FD23	INC IY
CB72	BIT 6, D	08	DEC BC	2C	INC L
CB73	BIT 6, E	0D	DEC C	33	INC SP
CB74	BIT 6, H	15	DEC D	EDAA	IND
CB75	BIT 6, L	1B	DEC DE	EDBA	INDR
CB7E	BIT 7, (HL)	1D	DEC E	EDA2	INI
DDCB057E	BIT 7, (IX + d)	25	DEC H	EDB2	INIR
FDCB057E	BIT 7, (IY + d)	2B	DEC HL	E9	JP (HL)
CB7F	BIT 7, A	DD28	DEC IX	DDE9	JP (IX)
CB78	BIT 7, B	FD2B	DEC IY	FDE9	JP (IY)
CB79	BIT 7, C	2D	DEC L	DA8405	JP C, NN
CB7A	BIT 7, D	3B	DEC SP	FA8405	JP M, NN
CB7B	BIT 7, E	F3	DI	D28405	JP NC, NN
CB7C	BIT 7, H	102E	DJNZ DIS	C38405	JP NN
CB7D	BIT 7, L	FB	EI	C28405	JP NZ, NN
DC8405	CALL C, NN	E3	EX (SP), HL	F28405	JP P, NN
FC8405	CALL M, NN	DDE3	EX (SP), IX	EA8405	JP PE, NN
D48405	CALL NC, NN	FDE3	EX (SP), IY	E28405	JP PO, NN
CD8405	CALL NN	08	EX AF, AF′	CA8405	JP Z, NN
C48405	CALL NZ, NN	EB	EX DE, HL	382E	JR C, DIS
F48405	CALL P, NN	D9	EXX	182E	JR DIS
EC8405	CALL PE, NN	76	HALT	302E	JR NC, DIS
E48405	CALL PO, NN	ED46	IM 0	202E	JR NZ, DIS
CC8405	CALL Z, NN	ED56	IM 1	282E	JR Z, DIS
3F	CCF	ED5E	IM 2	02	LD (BC), A
BE	CP (HL)	ED78	IN A, (C)	12	LD (DE), A
DDBE05	CP (IX + d)	DB20	IN A, (N)	77	LD (HL), A
FDBE06	CP (IY + d)	ED40	IN B, (C)	70	LD (HL), B

Z80 INSTRUCTIONS SORTED BY MNEMONIC / Appendix B

OBJ CODE	SOURCE STATEMENT	OBJ CODE	SOURCE STATEMENT	OBJ CODE	SOURCE STATEMENT
71	LD (HL), C	FD4605	LD B, (IY + d)	66	LD H, (HL)
72	LD (HL), D	47	LD B, A	DD6605	LD H, (IX + d)
73	LD (HL), E	40	LD B, B	FD6606	LD H, (IY + d)
74	LD (HL), H	41	LD B, C	67	LD H, A
75	LD (HL), L	42	LD B, D	60	LD H, B
3620	LD (HL), N	43	LD B, E	61	LD H, C
DD7705	LD (IX + d), A	44	LD B, H, NN	62	LD H, D
DD7005	LD (IX + d), B	45	LD B, L	63	LD H, E
DD7105	LD (IX + d), C	0620	LD B, N	64	LD H, H
DD7205	LD (IX + d), D	ED4B8405	LD BC, (NN)	65	LD H, L
DD7305	LD (IX + d), E	018405	LD BC, NN	2620	LD H, N
DD7405	LD (IX + d), H	4E	LD C, (HL)	2A8405	LD HL, (NN)
DD7505	LD (IX + d), L	DD4E05	LD C, (IX + d)	218405	LD HL, NN
DD360520	LD (IX + d), N	FD4E05	LD C, (IY + d)	ED47	LD I, A
FD7705	LD (IY + d), A	4F	LD C, A	DD2AB405	LD IX, (NN)
FD7005	LD (IY + d), B	48	LD C, B	DD218405	LD IX, NN
FD7105	LD (IY + d), C	49	LD C, C	FD2A8405	LD IY, (NN)
FD7205	LD (IY + d), D	4A	LD C, D	FD218405	LD IY, NN
FD7305	LD (IY + d), E	4B	LD C, E	6E	LD L, (HL)
FD7405	LD (IY + d), H	4C	LD C, H	DD6E05	LD L, (IX + d)
FD7505	LD (IY + d), L	4D	LD C, L	FD6E05	LD L, (IY + d)
FD360520	LD (IY + d), N	0E20	LD C, N	6F	LD L, A
328405	LD (NN), A	56	LD D, (HL)	68	LD L, B
ED438405	LD (NN), BC	DD5605	LD D, (IX + d)	69	LD L, C
ED538405	LD (NN), DE	FD5605	LD D, (IY + d)	6A	LD L, D
228405	LD (NN), HL	57	LD D, A	6B	LD L, E
DD228405	LD (NN), IX	50	LD D, B	6C	LD L, H
FD228405	LD (NN), IY	51	LD D, C	6D	LD L, L
ED738405	LD (NN), SP	52	LD D, D	2E20	LD L, N
0A	LD A, (BC)	53	LD D, E	ED788405	LD SP, (NN)
1A	LD A, (DE)	54	LD D, H	F9	LD SP, HL
7E	LD A, (HL)	55	LD D, L	DDF9	LD SP, IX
DD7E05	LD A, (IX + d)	1620	LD D, N	FDF9	LD SP, IY
FD7E05	LD A, (IY + d)	ED5B8405	LD DE, (NN)	318405	LD SP, NN
3A8405	LD A, (NN)	118405	LD DE, NN	EDA8	LDD
7F	LD A, A	5E	LD E, (HL)	ED88	LDDR
78	LD A, B	DD5E05	LD E, (IX + d)	EDA0	LDI
79	LD A, C	FD5E05	LD E, (IY + d)	EDB0	LDIR
7A	LD A, D	5F	LD E, A	ED44	NEG
7B	LD A, E	58	LD E, B	00	NOP
7C	LD A, H	59	LD E, C	B6	OR (HL)
ED57	LD A, I	5A	LD E, D	DDB605	OR (IX + d)
7D	LD A, L	5B	LD E, E	FDB605	OR (IY + d)
3E20	LD A, N	5C	LD E, H	B7	OR A
46	LD B, (HL)	5D	LD E, L	B0	OR B
DD4605	LD B, (IX + d)	1E20	LD E, N	B1	OR C

Z80 INSTRUCTIONS SORTED BY MNEMONIC / Appendix B

OBJ CODE	SOURCE STATEMENT	OBJ CODE	SOURCE STATEMENT	OBJ CODE	SOURCE STATEMENT
B2	OR D	CB8B	RES 1, E	CBB7	RES 6, A
B3	OR E	CB8C	RES 1, H	CBB0	RES 6, B
B4	OR H	CB8D	RES 1, L	CBB1	RES 6, C
B5	OR L	CB96	RES 2, (HL)	CBB2	RES 6, D
F620	OR N	DDCB0596	RES 2, (IX + d)	CBB3	RES 6, E
EDBB	OTDR	FDCB0596	RES 2, (IY + d)	CBB4	RES 6, H
EDB3	OTIR	CB97	RES 2, A	CBB5	RES 6, L
ED79	OUT (C), A	CB90	RES 2, B	CBBE	RES 7, (HL)
ED41	OUT (C), B	CB91	RES 2, C	DDCB05BE	RES 7, (IX + d)
ED49	OUT (C), C	CB92	RES 2, D	FDCB05BE	RES 7, (IX + d)
ED51	OUT (C), D	CB93	RES 2, E	CBBF	RES 7, A
ED59	OUT (C), E	CB94	RES 2, H	CBB8	RES 7, B
ED61	OUT (C), H	CB95	RES 2, L	CBB9	RES 7, C
ED69	OUT (C), L	CB9E	RES 3, (HL)	CBBA	RES 7, D
D320	OUT (N), A	DDCB059E	RES 3, (IX + d)	CBBB	RES 7, E
EDAB	OUTD	FDCB059E	RES 3, (IY + d)	CBBC	RES 7, H
EDA3	OUTI	CB9F	RES 3, A	CBBD	RES 7, L
F1	POP AF	CB98	RES 3, B	C9	RET
C1	POP BC	CB99	RES 3, C	D8	RET C
D1	POP DE	CB9A	RES 3, D	F8	RET M
E1	POP HL	CB9B	RES 3, E	D0	RET NC
DDE1	POP IX	CB9C	RES 3, H	C0	RET NZ
FDE1	POP IY	CB9D	RES 3, L	F0	RET P
F5	PUSH AF	CBA6	RES 4, (HL)	E8	RET PE
C5	PUSH BC	DDCB05AB	RES 4, (IX + d)	E0	RET PO
D5	PUSH DE	FDCB05AB	RES 4, (IY + d)	C8	RET Z
E5	PUSH HL	CBA7	RES 4, A	ED4D	RETI
DDE5	PUSH IX	CBA0	RES 4, B	ED45	RETN
FDE5	PUSH IY	CBA1	RES 4, C	CB16	RL (HL)
CB86	RES 0, (HL)	CBA2	RES 4, D	DDCB0516	RL (IX + d)
DDCB0586	RES 0, (IX + d)	CBA3	RES 4, E	FDCB0516	RL (IY + d)
FDCB0586	RES 0, (IY + d)	CBA4	RES 4, H	CB17	RL A
CB87	RES 0, A	CBA5	RES 4, L	CB10	RL B
CB80	RES 0, B	CBAE	RES 5, (HL)	CB11	RL C
CB81	RES 0, C	DDCB05AE	RES 5, (IX + d)	CB12	RL D
CB82	RES 0, D	FDCB05AE	RES 5, (IY + d)	CB13	RL E
CB83	RES 0, E	CBAF	RES 5, A	CB14	RL H
CB84	RES 0, H	CBA8	RES 5, B	CB15	RL L
CB85	RES 0, L	CBA9	RES 5, C	17	RLA
CB8E	RES 1, (HL)	CBAA	RES 5, D	CB06	RLC (HL)
DDCB058E	RES 1, (IX + d)	CBAB	RES 5, E	DDCB0506	RLC (IX + d)
FDCB058E	RES 1, (IY + d)	CBAC	RES 5, H	FDCB0506	RLC (IY + d)
CB8F	RES 1, A	CBAD	RES 5, L	CB07	RLC A
CB88	RES 1, B	CBB6	RES 6, (HL)	CB00	RLC B
CB89	RES 1, C	DDCB05B6	RES 6, (IX + d)	CB01	RLC C
CB8A	RES 1, D	FDCB05B6	RES 6, (IY + d)	CB02	RLC D

Z80 INSTRUCTIONS SORTED BY MNEMONIC / Appendix B

OBJ CODE	SOURCE STATEMENT	OBJ CODE	SOURCE STATEMENT	OBJ CODE	SOURCE STATEMENT
CB03	RLC E	DE20	SBC A, N	CBE6	SET 4, (HL)
CB04	RLC H	ED42	SBC HL, BC	DDCB05E6	SET 4, (IX + d)
CB05	RLC L	ED52	SBC HL, DE	FDCB05E6	SET 4, (IY + d)
07	RLCA	ED62	SBC HL, HL	CBE7	SET 4, A
ED6F	RLD	ED72	SBC HL, SP	CBE0	SET 4, B
CB1E	RR (HL)	37	SCF	CBE1	SET 4, C
DDCB051E	RR (IX + d)	CBC6	SET 0, (HL)	CBE2	SET 4, D
FDCB051E	RR (IY + d)	DDCB05C6	SET 0, (IX + d)	CBE3	SET 4, E
CB1F	RR A	FDCB05C6	SET 0, (IY + d)	CBE4	SET 4, H
CB18	RR B	CBC7	SET 0, A	CBE5	SET 4, L
CB19	RR C	CBC0	SET 0, B	CBEE	SET 5, (HL)
CB1A	RR D	CBC1	SET 0, C	DDCB05EE	SET 5, (IX + d)
CB1B	RR E	CBC2	SET 0, D	FDCB05EE	SET 5, (IY + d)
CB1C	RR H	CBC3	SET 0, E	CBEF	SET 5, A
CB1D	RR L	CBC4	SET 0, H	CBE8	SET 5, B
1F	RRA	CBC5	SET 0, L	CBE9	SET 5, C
CB0E	RRC (HL)	CBCE	SET 1, (HL)	CBEA	SET 5, D
DDCB050E	RRC (IX + d)	DDCB05CE	SET 1, (IX + d)	CBEB	SET 5, E
FDCB050E	RRC (IY + d)	FDCB05CE	SET 1, (IY + d)	CBEC	SET 5, H
CB0F	RRC A	CBCF	SET 1, A	CBED	SET 5, L
CB08	RRC B	CBC8	SET 1, B	CBF6	SET 6, (HL)
CB09	RRC C	CBC9	SET 1, C	DDCB05F6	SET 6, (IX + d)
CB0A	RRC D	CBCA	SET 1, D	FDCB05F6	SET 6, (IY + d)
CB0B	RRC E	CBCB	SET 1, E	CBF7	SET 6, A
CB0C	RRC H	CBCC	SET 1, H	CBF0	SET 6, B
CB0D	RRC L	CBCD	SET 1, L	CBF1	SET 6, C
0F	RRCA	CBD6	SET 2, (HL)	CBF2	SET 6, D
ED67	RRD	DDCB05D6	SET 2, (IX + d)	CBF3	SET 6, E
C7	RST 0	FDCB05D6	SET 2, (IY + d)	CBF4	SET 6, H
D7	RST 10H	CBD7	SET 2, A	CBF5	SET 6, L
DF	RST 18H	CBD0	SET 2, B	CBFE	SET 7, (HL)
E7	RST 20H	CBD1	SET 2, C	DDCB05FE	SET 7, (IX + d)
EF	RST 28H	CBD2	SET 2, D	FDCB05FE	SET 7, (IY + d)
F7	RST 30H	CBD3	SET 2, E	CBFF	SET 7, A
FF	RST 38H	CBD4	SET 2, H	CBF8	SET 7, B
CF	RST 8	CBD5	SET 2, L	CBF9	SET 7, C
9E	SBC A, (HL)	CBD8	SET 3, B	CBFA	SET 7, D
DD9E05	SBC A, (IX + d)	CBDE	SET 3, (HL)	CBFB	SET 7, E
FD9E05	SBC A, (IY + d)	DDCB05DE	SET 3, (IX + d)	CBFC	SET 7, H
9F	SBC A, A	FDCB05DE	SET 3, (IY + d)	CBFD	SET 7, L
98	SBC A, B	CBDF	SET 3, A	CB26	SLA (HL)
99	SBC A, C	CBD9	SET 3, C	DDCB0526	SLA (IX + d)
9A	SBC A, D	CBDA	SET 3, D	FDCB0526	SLA (IY + d)
9B	SBC A, E	CBDB	SET 3, E	CB27	SLA A
9C	SBC A, H	CBDC	SET 3, H	CB20	SLA B
9D	SBC A, L	CBDD	SET 3, L	CB21	SLA C

Z80 INSTRUCTIONS SORTED BY MNEMONIC / Appendix B

OBJ CODE	SOURCE STATEMENT	OBJ CODE	SOURCE STATEMENT	OBJ CODE	SOURCE STATEMENT
CB22	SLA D	FDCB053E	SRL (IY + d)	93	SUB E
CB23	SLA E	CB3F	SRL A	94	SUB H
CB24	SLA H	CB38	SRL B	95	SUB L
CB25	SLA L	CB39	SRL C	D620	SUB N
CB2E	SRA (HL)	CB3A	SRL D	AE	XOR (HL)
DDCB052E	SRA (IX + d)	CB3B	SRL E	DDAE05	XOR (IX + d)
FDCB052E	SRA (IY + d)	CB3C	SRL H	FDAE05	XOR (IY + d)
CB2F	SRA A	CB3D	SRL L	AF	XOR A
CB28	SRA B	96	SUB (HL)	A8	XOR B
CB29	SRA C	DD9605	SUB (IX + d)	A9	XOR C
CB2A	SRA D	FD9605	SUB (IY + d)	AA	XOR D
CB2B	SRA E	97	SUB A	AB	XOR E
CB2C	SRA H	90	SUB B	AC	XOR H
CB2D	SRA L	91	SUB C	AD	XOR L
CB3E	SRL (HL)	92	SUB D	EE20	XOR N
DDCB053E	SRL (IX + d)				

Appendix C
Z80/8080 Instruction Equivalency

8080/Z80 INSTRUCTION EQUIVALENCY (SAME OP-CODES)

Eight-bit load group

8080	Z80
MOV	LD (all combinations)
MVI	LD
LDA	LD A, (nn)
STA	LD (nn), A
LDAX	LD LD A, (zz)
LDAI	LD A, I
LDAR	LD A, r
STAI	LD I, A
STAR	LD r, A

Sixteen-bit load group

8080	Z80
LXI	LD rr, nn
LBCD	LD BC, (nn)
LDED	LD DE, (nn)
LHLD	LD HL, (nn)
LIXD	LD IX, (nn)
LIYD	LD IY, (nn)
LSPD	LD SP, (nn)
SBCD	LD (nn), BC
SDED	LD (nn), DE
SHLD	LD (nn), HL
SIXD	LD (nn), IX
SIYD	LD (nn), IY
SSPD	LD (nn), SP
SPHL	LD (nn), HL
SPIX	LD (nn), IX
SPIY	LD (nn), IY
PUSH	PUSH (all mnemonics)
POP	POP (all mnemonics)

Z80/8080 INSTRUCTION EQUIVALENCY / Appendix C

Exchange, Transfer, Search Group

8080	Z80
XCHG	EX DE,HL
EXAF	EX AF,AF'
EXX	EXX
XTHL	EX (SP),HL
XTIX	EX (SP),IX
XTIY	EX (SP),IY
LDI	LDI
LDIR	LDIR
LDD	LDD
LDDR	LDDR
CCI	CPI
CCIR	CPIR
CCD	CPD
CCDR	CPDR

Eight-bit Arithmetic/Logical Group

8080	Z80
ADD	ADD (all combinations)
ADI	ADD A,n
ADC	ADC
ACI	ADC A,n
SUB	SUB
SUI	SUB A,n
SBC	SBC
SBI	SBC A,n
ANA	AND
ANI	AND A,n
ORA	OR
ORI	OR A,n
XRA	XOR
XRI	XOR A,n
CMP	CP
CPI	CP A,n

INR	INC r
INR M	INC (HL)
INR d(ii)	INC (Iii+d)
DCR r	DEC r
DCR M	DEC (HL)
DCR d(ii)	DEC (Iii+d)

General-purpose Arithmetic/Control Group

8080	Z80
DAA	DAA
CMA	CPL
NEG	NEG
CMC	CCF
STC	SCF
NOP	NOP
DI	DI
HALT	HALT
EI	EI
IMØ	IMØ
IM1	IM1
IM2	IM2

Rotate and Shift Group

8080	Z80
DAD	ADD
DADC	ADC
DSBC	SBC HL,rr
DADX	ADD IX,tt
DADY	ADD IY,tt
INX rr	INC rr
INX ii	INC ii
DCX rr	DEC rr
DCX ii	DEC ii
RLC	RLCA
RAL	RLA
RRC	RRCA

Z80/8080 INSTRUCTION EQUIVALENCY / Appendix C

8080	Z80
RAR	RRA
RLCR r	RLC r
RLCR M	RLC (HL)
RLCR d(ii)	RLC (Iii+d)
RALR	RL
RRCR	RRC s
RARR	RR s
SLAR	SLA
SRAR	SRA
SRLR	SRL
RLD	RLD
RRD	RRD

Bit Test, Bit Set, Bit Reset Group

8080	Z80
BIT b,r	BIT b,r
BIT b,M	BIT b, (HL)
BIT b,d(ii)	BIT b,d(ii)
SET b,r	SET b,r
SET b,M	SET b, (HL)
SET b,d(ii)	SET b,d(ii)
RES b,s	RES b,s

Jump Group

8080	Z80
JMP	JP
JZ	JP Z
JNZ	JP NZ,nn
JC	JP C,nn
JNC	JP NC,nn
JPO	JP PO,nn
JPE	JP PE,nn
JP	JP P,nn
JM	JP M,nn
JO	JP PE,nn

JNO	JP PO,nn
JMPR	JR,e
JRZ	JR Z,e
JRNZ	JR NZ,e
JRC	JR C,e
JRNC	JR NC,e
DJNZ	DJNZ e
PCHL	JP (HL)
PCIX	JP (IX)
PCIY	JP (IY)

Call/Return Group

8080	Z80
Call nn	Call nn
CZ nn	CALL Z,nn
CNZ	CALL NZ,nn
CC nn	CALL C,nn
CNC	CALL NC
CPO	CALL PO
CPE	CALL PR
CP	CALL P
CM	CALL M,nn
CO	CALL PE,nn
CNO	CALL PO,nn
RET	RET
RZ	RET Z
RNZ	RET NZ
RC	RET C
RNC	RET NC
RPO	RET PO
RPE	RET PE
RP	RET P
RM	RET M
RO	RET PE
RNO	RET PO
RETI	RETI

8080	Z80
RETN	RETN
RST n	RST n

Input/Output Group

8080	Z80
IN n	IN A, (n)
INP r	IN r, (C)
INI	INI
INIR	INIR
INDR	INDR
OUT n	OUT (n), A
OUTP	OUT (C), r
OUTI	OUTI
OUTIR	OTIR
OUTD	OUTD
OUTDR	OTDR

Not all of the Z80 instructions have equivalents in the 8080 system. Those listed above use the same op-codes, so they can be plugged into either type of microcomputer. The principal difference between the software of the two different types of uP chip lies in the *timing*. 8080 software will execute on Z80 machines unless timing is important.

Index

Accumulator, 7
AC power line, 53
Adders, 37
Addition, 37
Address block decoding, 83
Addressing, in microcomputers, 78
A/D (*see* Analog-to-digital converter)
AIM-65, 9
ALU (*see* Arithmetic logic unit)
American Standard Code for Information Interchange (*see also* ASCII), 43
Analog, 2
Analog-to-digital conversion, 248ff
Analog-to-digital converter, 72
   types, 249
   binary ramp, 249ff
   dual slope integrator, 256
   flash, 254
   parallel, 254
   servo, 249ff
   successive approximation, 251ff
Analog-vs-digital, 2
Analog reference circuits, 209–222
Analog switches, 200
A register, 7
Arithmetic logic unit, 7, 24, 37
Arithmetic operations, 37
Artifact, recirculation, 77
ASCII, 9, 43
Asynchronous operation, 59

Band gap zener diode, 215
Bank select, 85
Base-10, 32
Binary coded decimal (BCD), 39
Binary (base-2) numbers, 34
Binary coded hexadecimal (BCH), 41
Binary coded octal (BCO), 39
Biquinary numbers, 32
Bus, address, 78
Bus drivers, 51
Byte directional mode, 54–55

Calibration error, 5
Character, 43
Chip enable, 50, 79
Code, character, 43
Code, excess-3, 42
Code, gray, 42
Code, machine, 42
Computer, cardiac output, 2, 77
Computer, mainframe, 14
Computer, programmable digital, 66
Computer, single-chip, 9
Constant current source, 256
Control signals, 88
Control systems, 1
Compensation, temperature, 156
CPU, 7, 22
Current loop, 119
Current reference sources, 220

DAC (see Digital-to-analog converter)
Data acquisition systems (DAS), 263-271
Data conversion—A/D, 248-262
Data conversion, software, 289
Data converters, 273-274
    D/A 240-247
Data creep (or "walk"), 269
Data direction registers, 68
Decimal number system, 32
Decoder, address, 78
Decoder, BCD-to-1-of-10, 83
Decoders, 8-bit, 79
Definitions, 7
Differential amplifiers, 144ff
    applications 150
Differential inputs, 134
Differentiation, 73
Differentiators, 153
Digital, 2
Digital-to-analog converter, 4, 94, 209
    output circuitry 280
Digitization error, 67
Direct memory access, 61, 99
Droop, 201

EEG evoked potentials computer, 286
End-of-conversion, 281
Error, amplifier gain, 5
Error, analog circuits, 5
Error, drift, 5
Error, reference potential, 5
Errors, sample & hold, 205
Excess-3 code, 42

Flag register, 24
Fourier series, 129

Gage factor, 166
Gray code, 42

Handshake, 55
Hexadecimal (base-16) numbers, 36
Hysteresis error, 5

Index registers, 23
Inductive kick, 49
Inertia error, 5
Integration, 72ff
Integrators, 152
Interfacing, 44ff
    A/D, 281
    current loop, 119

Interfacing (Contd.)
    DAC, 274
    data converters, 272-288
    IC logic families, 45
    I/O, 98-125
    I/O-based, 274
    keyboards, switches, and displays, 223-239
    LEDs, lamps, and relays, 47
    memory, 88
    RS-232, 119
Inverting follower, 135
I/O, 68, 81
I/O ports, 98
    hardware, 110
    serial, 108
    parallel, 105
Instrument, calculating, 2
Instruments, 1
Instruments, electronic, classification of, 1
Instrumentation, 66
Instrumentation engineer, 5
Interrupt vector, 22

Keyboard, 9, 224
KIM-1, 9
Kluge board, 17

Ladder circuit, binary weighted resistor, 241
Ladder circuit, R-2R, 244
Lamp driver, 49
Latch, 96, 103
Linear variable differential transformer, 181
Logarithmic amplifiers, 155
LVDT (see Linear variable differential transformer)

Microcomputer, 7, 8
Microprocessor, 7
Memory, dynamic, 91
Memory-mapping, 94, 99, 276, 293
Memory refresh, 23
Microprocessor, 66
Microprocessor, adv. of, 15
Microprocessor support chips, 54ff
Microcomputer, 11
Microcomputers, adv. of, 15
Modes, Z80-PIO, 55

Noninverting followers, 137
Noise, 130
Numbers systems, other than base-10, 33

INDEX

Octal (base-8) numbers 35
Offset current, 5
Offset voltage, 5
Operational amplifiers, 132–159
Operational amplifiers, power supplies for, 139
Operational amplifiers
  practical circuit, 147
  problems, 141
  ideal properties, 133
Open-collector TTL, 48
Optoisolators, 53
Output port, parallel, 103

Parallax error, 5
Peripherals, 9
Phototransistor, 53
Piezoresistivity, 163
Precision reference supplies, 212ff
Problem solving with microprocessor, 66
Program counter, 23, 29

Quantization error, 66–67

RAM (*see* Random access memory)
Ramp A/D converter, software, 290
Random access memory, 8
Read only memory, 89, 224, 290
Reference voltage ICs, 219ff
Relay driver, 49
ROM (*see* Read only memory)
RS-232, 119

Sample-hold circuits, 199–209
SBC (*see* Single board computer)
Schmitt trigger, 256
Select pulse, generation of, 99
Select signal, 81
Sensitivity, strain gage, 165
Sensitivity, transducer, 171
Signal, digital, 3
Signals & noise, 126–131
Single board computer, 9
Software, 5
Split-octal, 40
Stack pointer, 23
Status register, 31
Strain gage, 162
Strain gage—bonded/unbonded, 166
Strobe, 96
Strobe signal, 224
Successive approximation (A/D) converter, 293

Superboard II, 9
Support chip, 54
SYM-1, 10
Synchronous binary operation, 59

Tachometers, 186
Temperature, 1
  transducers for, 174
  transducers, semiconductor, 177
Thermistor, 156, 174
Thermocouples, 177
Touchtone®, 9
Transducers, 161–198
  definition, 162
  capacitive, 191
  fluid pressure, 188
  force and pressure, 187
  inductive, 180
  light, 190
  position/displacement, 183
  temperature, 174
  velocity/accelleration, 185
Transduction, 161
Trainer, 9
Tri-state, 50
Two's complement, 38

Universal asynchronous receiver/transmitter (UART), 111ff

Voltage controlled oscillator (VCO), 255
Voltage-to-frequency converter (V/F or VFC), 255

Wait states, 92
Weighted numbers systems, 33
Wheatstone bridge, 150, 162, 169

Zener diode, 210
Z80, 19ff
  pinouts, 25
Z80-CTC, 55, 64
Z80-DMA, 55, 61
Z80-PIO, 55
Z80-SIO, 55, 57

1-LSB, 210
1-LSB error, 5
1's complement, 38
6502, 10ff, 19, 28ff
2's complement, 38
10's complement, 38